TRACTOR
SUPERSTARS

THE GREATEST TRACTORS
OF ALL TIME

THARRAN E. GAINES

Quarto is the authority on a wide range of topics.

Quarto educates, entertains and enriches the lives of
our readers—enthusiasts and lovers of hands-on living.

www.quartoknows.com

First published in 2015 by Voyageur Press, an imprint of Quarto
Publishing Group USA Inc., 400 First Avenue North, Suite 400,
Minneapolis, MN 55401 USA

quartoknows.com
Visit our blogs at quartoknows.com

Voyageur Press titles are also available at discounts in bulk quantity
for industrial or sales-promotional use. For details write to Special
Sales Manager at Quarto Publishing Group USA Inc., 400 First
Avenue North, Suite 400, Minneapolis, MN 55401 USA.

To find out more about our books, visit us online at
www.quartoknows.com.

ISBN: 978-0-7603-4931-1

10 9 8 7 6 5 4 3 2

Library of Congress Cataloging-in-Publication Data

Gaines, Tharran E., 1950-
  Tractor superstars : the greatest tractors of all time / by Tharran
Gaines.
    pages cm
  ISBN 978-0-7603-4931-1 (paperback)
 1. Farm tractors. I. Title.
 TL233.G27 2015
 629.225'2--dc23
                                    2015021606

Acquiring Editor: Todd R. Berger
Project Manager: Caitlin Fultz
Art Director: Brad Springer
Cover Designer: Kent Jensen
Designer: Brad Norr
Layout: John Sticha

On the front cover: *Ralph W. Sanders*
On the back cover: *Ralph W. Sanders*
On the frontis: *Ralph W. Sanders*
On the title page: *Ralph W. Sanders*
On the contents page: *Ralph W. Sanders*

Printed in China

Contents

ACKNOWLEDGMENTS

Even though I have been writing about tractors and agricultural topics for nearly 40 years, I'll be the first to admit that there is still a lot about tractors that I don't know. So I have to credit a lot of people and sources for information in this book. Tractors, in general, have been part of my life since I was a child growing up in north-central Kansas. So, I first need to credit my late parents, Francis and Ada Gaines, for exposing me to tractors and agriculture from day one. As a result, my own list of tractor superstars wouldn't be complete without a Case Model D. Although the D isn't featured as a superstar in this book, it was built in the same era as the S/SC highlighted in Chapter 1: First in the Field. Equipped with high fenders and a hand clutch, it was the ideal model for a first-time tractor driver and the first tractor I ever drove. Other tractors that I fondly remember, that do appear in this book, include a Farmall C, a Farmall 560—which is the tractor I spent more hours on than any other—and an Allis-Chalmers 190XT.

I also owe a great deal of appreciation to my wife, Barb. Not only was she overly patient while I tried to fit this book in between all my other projects—ignoring a number of jobs around the house and in the yard in the process—but she spent several hours proofreading the copy and the photo captions for typographical and grammatical errors.

Others I wish to thank include Dave Mowitz, editor of *Ageless Iron* magazine, and Sherry Schaefer, editor of *Heritage Iron* magazine, for allowing me to reuse some of the information I acquired while researching articles for their respective publications. I also owe a debt of gratitude to the late Norm Swinford, author of *The Proud Heritage of AGCO Tractors* and *Allis-Chalmers Farm Equipment 1914–1985*, from which I acquired a wealth of data on Allis-Chalmers, White, Minneapolis-Moline, Oliver, Cockshutt, Hart-Parr, and Rumely tractors. Thanks also go to Jason Hasert, who authors an online magazine called *Toy Tractor Times*, as well as a website titled Big Tractor Power at www.bigtractorpower.com.

I also owe a debt of gratitude to Peter Easterlund, owner of the website TractorData, found at www.tractordata.com. While there is no guarantee of accuracy, TractorData has more specifications on more tractors than any other source I have found. And most of the tractor specifications found there agree with other sources I've been able to locate—if I've been able to locate any at all on some of the rare tractors. The other good source for specifications is the website for the Nebraska Tractor Test Laboratory at tractortestlab.unl.edu.

Of course, there are also a number of individuals whom I credit for expanding my knowledge of classic tractors over the years, particularly those who contributed information for feature articles I've written or for the tractor restoration books I've written for Voyageur Press. Among them are Roy Ritter and Estel Theis, John Deere two-cylinder tractor enthusiasts who have since passed away, taking a wealth of information with them. Other sources include Clint and Stan Stamm, farmers and Minneapolis-Moline enthusiasts from Washington, Kansas; Jeff Gravert, a Cockshutt tractor enthusiast from Clay Center, Nebraska; Larry Karg, a farmer and Allis-Chalmers collector from Hutchinson, Minnesota; Eugene, Gaylen, and Martin Mohr, as well as their late father, Roger Mohr, collectors of Minneapolis-Moline tractors from Vail, Iowa; and Tim Brannon, owner of B & G Equipment, an AGCO brand dealer and longtime Allis-Chalmers enthusiast in Paris, Tennessee.

Last of all, though, I want to thank Todd R. Berger, acquiring editor with Voyageur Press, for his guidance and direction on this project and for his patience when I found myself running behind schedule.

INTRODUCTION

What is a tractor superstar? Well, that's hard for any one person to say. A tractor superstar can easily be different things to different people, depending to a great extent on when you were born. To those of us who are baby boomers, it is the tractors of the 1960s and 1970s that are most remembered as something special. That was the era in which tractor power broke the 100-horsepower barrier. It was also the time period in which cabs on tractors became standard equipment, turbochargers were added to increase power, and big bore engines cranked out new levels of torque. I can still remember when a neighbor in my home state of Kansas bought a new John Deere 5010. To me, a teenager running a Farmall 560 on my side of the road, that new green machine was a monster. And the Allis-Chalmers D21, which pushed Allis-Chalmers over the 100-horsepower limit, is still my favorite tractor of all time. I don't know if it is the power, the styling, or the chrome that has me hooked.

However, if you are currently in your 80s, a superstar tractor to you might be the first one that entered the market with a live power take-off (PTO), rubber tires instead of steel wheels, or a three-point hitch. Whether it was a success at the time or not, a tractor that was the first to introduce a new timesaving or efficiency feature is often remembered as a superstar in its era. If we were to feature tractors worldwide or consider today's tractors, we would have to include Fendt for being the first to introduce a tractor with a continuously variable transmission (CVT). We would also have to do some research to find out who was the first to offer satellite-assisted guidance.

There will always be those who debate who was first in certain categories, regardless of the era in which the innovation occurred. Take the independent PTO, for example. While the Oliver 88 is often credited with being the first mass-produced US-built tractor with an independent or live PTO—which continues to operate when the operator pushes in the clutch to stop ground travel—there's still a great deal of debate on the subject. Minneapolis-Moline is said to have developed an independent PTO, but it was never marketed. However, Cockshutt was the first to receive the credit when the company tested the Model 30 at the Nebraska Tractor Test Laboratory. So Cockshutt is the one that gets the credit in Chapter 1: First in the Field.

Of course, if you grew up farming with horses, a superstar might be the first tractor you ever owned—even if you owned a number of bigger, more powerful tractors after that. Those first tractors accounted for a lot of machines, too, as evident in Chapter 8: Record Sales. In fact, there have been fewer than a handful of tractor models that have exceeded 200,000 units since the 1930s. Even though some of the best-selling tractors of all times were introduced and/or marketed during the Great Depression, they represented an opportunity to dramatically reduce labor, expand the operation, or simply live a better life. So anyone who could afford one was trading in horses or mules for a tractor.

A change from the norm always caught people's attention, too. Whether it was something different from anything else in the industry or something unique within a brand—like John Deere's dramatic switch from two-cylinder engines to four- and six-cylinder engines—it was often enough to push a tractor into superstar status. And who can ignore a boost in power? Since the early days of farming, farmers have held contests to see whose horse could pull the biggest loads. So it comes as no surprise that farmers and manufacturers were always on a quest for the most powerful tractor. The irony is that we have lawn tractors today that boast more horsepower than some of the early farm tractors.

Of course, 20 or 30 years from now there may be a whole new list of tractor superstars. No doubt it will include the first remote-control tractor to perform without an operator. And maybe the list will include the first tractor to have over 1,000 horsepower. German-based Fendt, which is part of AGCO Corporation in Duluth, Georgia, has already introduced the industry's first standard rigid-frame tractor with 500 engine horsepower. So nothing is impossible when you consider how far the tractor industry has come in the last 100 years. Whether it earned the distinction through power, innovation, or a dramatic change in design, the superstars of the past have simply become the steppingstones to the future.

First
IN THE
Field

There's always something to be said about being the first to introduce an innovation. Sometimes it means being remembered in history, like Wilbur and Orville Wright after they became the first to build an airplane that actually flew. Other innovations were simply ahead of their time and took a few more years to catch on. It was no different in the tractor industry.

Some firsts, like the first three-point hitch developed by Harry Ferguson and the first tractor with an independent PTO, are still remembered. Those were also some of the innovations that caught on immediately. The benefits were so obvious to potential customers that other tractor manufacturers were practically forced to adopt the same technology or develop a similar design if they wanted a share in the market.

Other firsts, like rubber tires, took a little persuasion before they were widely accepted. Who could believe that rubber would provide better traction in the field than steel lugs that dug into the dirt? It was something people had to see for themselves to believe. Of course, other revolutionary ideas—like tractor cabs and front-wheel assist—were simply ahead of their time. Sometimes it wasn't even the tractor manufacturer or a lack of sales that was at fault. In this chapter, you'll note that the Caterpillar Challenger is recognized as the first track tractor with rubber tracks. But maybe it should be said that it was the first successful tractor with rubber tracks. In reality, Cletrac introduced a crawler with rubber tracks much earlier, in 1940, making it the first in the field with such a unique feature. Unfortunately, the technology of the era could not produce a stable rubber track that didn't stretch during use. So Cletrac went back to steel tracks before the idea even got off the ground.

It wasn't always the big things that made a significant impact on customers, either. Sometimes it was something as simple as being the first tractor manufacturer to use a dry, replaceable-element air cleaner in place of the oil-bath air cleaner. Not only could it go for long periods without cleaning, but even when cleaning did become necessary, it wasn't nearly the dirty, messy job required to dump and replace the oil in the oil-bath container.

The pages that follow represent only a few of the most noteworthy firsts in the tractor industry. There have obviously been hundreds of innovations that

Ralph W. Sanders

have been introduced between the time farm tractors were developed and today's era of tractors that have everything from satellite-guided steering to cab suspension. But with every innovation that was introduced, somebody had to be first.

Hart-Parr 17-30

Years of production	1903–1906
Engine	Hart-Parr, 1,654 cubic inches (27.1 liters), 2 cylinders
Fuel	Gasoline
Horsepower	30 belt (claimed), 17 drawbar (claimed)
Rated RPM	270
Transmission	Gear
Weight	14,500 pounds
Original price	Unknown

Charles W. Hart and Charles H. Parr were still engineering students at the University of Wisconsin in Madison when they founded the Hart-Parr Gasoline Engine Company in 1897 and began producing stationary engines. However, they soon outgrew their operation in Madison and moved the business to Charles City, Iowa, where they expanded into tractor manufacturing in 1901.

Although the 17-30 wasn't the first tractor to come out of the new plant, it was one of the first round of tractors the company produced; it was historical in that it was one of the first tractors to actually be called a "tractor." Up to this point, most units had been called "traction engines." Hart and Parr combined the words "traction" and "power" and their Latin roots to create the term "tractor." The Model 17-30 was also responsible for Hart-Parr being recognized as the first true tractor manufacturer in the industry, when more than a dozen units were factory-assembled in 1903 alone.

The very first Hart-Parr, or Model No. 1, also claimed 17 drawbar and 30 belt horsepower, hence the model number 17-30 for the later version. However, the No. 1 was essentially one of the company's stationary engines geared to the rear wheels on a steerable chassis. Still, it was noted as being the world's first production farm tractor with an internal combustion engine. The No. 2 followed with a more complete engine/transmission powertrain, but no seat for the operator. No. 3 in the series featured the same horizontal two-cylinder engine and is, today, on display in the Smithsonian Institution as one of the first tractors available to American farmers.

Introduced in 1903, the 17-30 was still based on the prior No. 3 model, which featured a huge 1,654-cubic-inch engine that utilized water injection to

The Hart-Parr 17-30 represented a number of "firsts" for the Hart-Parr Gasoline Engine Company. It was the first tractor to come out of a new tractor plant and one of the first to be called a "tractor." *Floyd County Museum, Charles City, Iowa*

control detonation. The early 17-30 models also featured a closed radiator for oil coolant, which depended on exhaust-induced draft for cooling. However, this idea was soon abandoned in favor of a corrugated fin radiator that appeared in all later models.

Pioneer 30-60

Based in Winona, Minnesota, the Pioneer Tractor Company was an early leader in the use of tractor cabs and sheet metal to enclose the entire engine and drivetrain. In fact, the 1910 Pioneer 30-60, which was one of the first tractors built by the company, had a canopy-covered "cab" complete with front and side windows and an upholstered seat. Pioneer also claimed to be

Years of production	1910–1927
Engine	Pioneer Tractor, 1,232 cubic inches (20.2 liters), 4 cylinders double-opposed engine
Fuel	Gasoline
Horsepower	60 belt (claimed), 30 drawbar (claimed)
Rated RPM	270
Transmission	3 speeds forward, 1 reverse
Weight	23,600 pounds
Original price	Approximately $4,000

first with other features in tractor design, including an entirely enclosed and oil-bathed drive gear train, all machine-cut steel gears, a three-speed sliding gear transmission, and an automotive-type front axle and steering system.

Built for breaking up the prairie with a plow and for road construction applications, the Pioneer 30-60 was big in every way, from its 7-inch piston bore to the 100-gallon fuel tank. The dimensions were equally impressive, considering a length of 20 feet 2½ inches, a width of 10 feet 10 inches, and steel rear wheels that measured 8 feet in diameter with a 24-inch face. The front wheels measured 5 feet in diameter with a 12-inch face.

It was claimed that the Pioneer 30-60, when new, could easily handle a 10-bottom plow (14-inch bottoms) and, in one Winnipeg tractor trial, a 30-60 did indeed pull a 10-bottom John Deere plow with ease in the plowing segment of

The Pioneer brand name is rather fitting for the 30-60, since the company was an early pioneer in the use of a tractor cab with front and side windows and an enclosed engine. It was also ahead of its time in power as it was claimed the 30-60 could easily pull a 10-bottom plow. *Ralph W. Sanders*

the contest. Unfortunately, Pioneer did have a drawback when it came to belt applications. Due to its design, the engine was arranged so the drive pulley and water pump worked together. Consequently, if the belt pulley had to be stopped, the water pump also quit, which could quickly lead to an overheated engine—particularly during hot summer months when most small-grain threshing took place.

Due to its success in breaking prairie sod, the Pioneer 30-60 quickly found success in Canada, as well as the Upper Midwest. However, Canadian tariffs on imported farm machinery added considerably to the cost of imported tractors. To compensate, Pioneer built a small factory at Calgary, Alberta, in 1912 with the idea of moving the manufacture of the 30-60s to Canada using components brought from Winona. In doing so, the company avoided the Canadian tariffs and consequently sold a number of 30-60s on the prairies, particularly in Saskatchewan.

Like so many tractor companies, though, Pioneer was not able to survive the move to smaller tractors. Pioneer underwent reorganization in the early 1920s but did not recover and disappeared as a company in 1927.

Holt 60

The 1911 Holt 60 was one of the first tractors with a gasoline engine built by Holt Manufacturing Company. Prior to this, Holt tractors—both wheeled and tracked—were powered by steam, which soon proved to be a problem in certain applications. As the tractors worked in the fields, they emitted hot cinders that often started fires, especially in wheat and hay fields that were being harvested. The steam boilers would also explode on occasion when steam pressure was inadvertently allowed to become too high or water levels in the boilers became too low.

Although the Holt 60 Caterpillar was originally built in Holt's Stockton, California, factory, the company began production of the Holt 40-60 Caterpillar in its East Peoria, Illinois, plant the same year. The primary difference between the two versions of the Holt 60 was in the head and valve

Years of production	1911–1917
Engine	Holt, 1,230 cubic inches (20 liters), 4 cylinders
Fuel	Gasoline
Horsepower	60 belt (claimed), 17 drawbar (claimed)
Rated RPM	1,650
Transmission	2 speeds forward, 1 reverse
Weight	22,000 pounds
Original price	$4,200

Part tractor and part crawler, the Holt 60 was the first of the Holt tractors—both wheeled and tracked—to feature a gasoline engine, which proved safer for agricultural applications than the earlier steam-powered version. *Ralph W. Sanders*

design of the engines. In the meantime, the Holt 60 still retained the tiller wheel followed by a pair of steel tracks, and steering was accomplished by releasing a one-track clutch and pivoting the front tiller wheel. Since there were no brakes, turns generally required a wide radius.

Not to be outdone, C. L. Best Gas Traction Company introduced its own Model 70 "Tracklayer" tractor in 1913. Of course, the model numbers at this time were based upon either horsepower (hence, the Holt 60) or the weight of the tractor (such as the Holt 10-Ton, which came later). Holt and Best continued as rivals both in the field and in the courts of law through litigation until the two companies finally consolidated as the Caterpillar Tractor Company in 1925.

Twin City 12-20

W hen it was introduced in 1919, the Twin City 12-20 featured more than a clean design with its compact unit-frame design. (That is, the engine and transmission were stressed units that formed the frame of the tractor.)

The 12-20 also featured an important engine feature that improved the engine's efficiency and gave it more power. The 340.5-cubic-inch engine in the 12-20 became the first to offer four valves per cylinder—two intake valves and two exhaust valves—permitting much greater capacity for air intake and exhaust movement than single intake and exhaust valves could attain.

Years of production	1919–1926
Engine	Twin City, 340.5 cubic inches (5.6 liters), 4 cylinders
Fuel	Kerosene
Horsepower	27.93 belt (tested), 18.43 drawbar (tested)
Rated RPM	1,000
Transmission	2 speeds forward, 1 reverse
Weight	5,000 pounds
Original price	Unknown

In addition to a unique unit-frame design—which means the engine and transmission formed the tractor frame—the Twin City 12-20 was the first to feature four valves per cylinder on its four-cylinder engine. *Ralph W. Sanders*

In an effort to offset its higher price, Twin City claimed that because of its special engine design and slower-rpm engine, their tractors would last at least three years longer than any competitive model, making it a lower-cost investment. *Ralph W. Sanders*

The design proved so successful that the 12-20 soon became the base for an even more powerful 17-28. Thanks in part to its rugged construction, four valves per cylinder, and a slow, 1,000-rpm engine, owners claimed lower cost per horsepower for more years. In fact, Minneapolis-Moline, which absorbed the Twin City line, claimed that its tractors would last at least three years longer than any competitive model, making it a lower-cost investment, despite its higher initial purchase price.

Years of production	1925–1928
Engine	Holt/Caterpillar, 251 cubic inches (4.1 liters), 4 cylinders
Fuel	Gasoline
Horsepower	25 belt (claimed), 15 drawbar (claimed)
Rated RPM	1,000
Transmission	3 speeds forward, 1 reverse
Weight	4,040 pounds
Original price	$1,850 (1927)

Caterpillar 2-Ton

The Caterpillar 2-Ton was a continuation of a tractor that had started life as the Holt T-35 in 1921. Unlike some previous track tractors, however, it was a small tractor that was aimed from the start at the agricultural, logging, and construction industries, instead of the construction field. Initially introduced at a price of just $375, the T-35 soon became the Holt 2-Ton to match the designations of other Holt tractors, including the 5-Ton and 10-Ton.

In 1925, the 2-Ton also became one of Holt's most successful contributions to the Caterpillar line during the merger with Best that followed hard times experienced by both former rivals. At the time of the consolidation, Holt was the larger partner. Hence, it provided three of the five tractors selected from the two brands to form the new Caterpillar line. One of these was the 2-Ton. Of course, the name Caterpillar also came from the Holt family of tractors. Not long after the first Holt was tested in 1904, a commentator reportedly

Early promotional ads for the Holt T-35, which soon became the 2-Ton, were clearly aimed at the agricultural market, stating that the "supreme small tractor" could be viewed in person at state and county fairs. *Ralph W. Sanders*

remarked, "She crawls along like a caterpillar." The name stuck and soon became the company trademark.

Once the 2-Ton became part of the Caterpillar brand, it continued for another three years without any mechanical changes. However, Caterpillar did make a few cosmetic changes with one of the most notable being the model name "2 TON" cast into the radiator side plates.

One unusual feature of the 2-Ton was that the three-speed transmission was located behind the rear axle, rather than in front of it, which is normal in most other tractors. The engine also utilized an overhead cam that was equally unusual at the time. In the meantime, all of the components on the tractor, except for the carburetor and magneto, were made by Holt, and later by Caterpillar.

Years of production	1934–1942
Engine	Hercules, 226 cubic inches (3.7 liters), 4 cylinders
Fuel	Gasoline
Horsepower	22.11 belt (claimed), 30.5 drawbar (claimed)
Rated RPM	1,300
Transmission	3 speeds forward, 1 reverse
Weight	5,200 pounds
Original price	$1,445

Cletrac EG

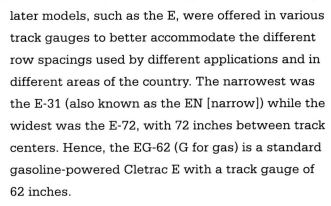

Founded by Clarence and Rollin White in 1916, the Cleveland Tractor Company—condensed in the model name to "Cletrac"—was focused from its inception on small and medium farm crawler tractors.

A unique innovation of Cletrac tractors was their differential steering called Tru-Traction, which was introduced on the company's first tractor—even though it didn't yet carry the trademark name. The conventional method of steering a tracked tractor was to declutch the inside track, which reduces the traction on that side, resulting in a turn. Instead, Cletrac introduced a planetary gear set that was controlled by a brake at each drive cog. That allowed the powertrain to transmit controlled amounts of power to each track, ensuring full traction at all times, even during turns.

The Cletrac EG-62, shown here, was just one of the early Cletrac models targeted at the ag market, rather than construction companies. In fact, many later models, such as the E, were offered in various track gauges to better accommodate the different row spacings used by different applications and in different areas of the country. The narrowest was the E-31 (also known as the EN [narrow]) while the widest was the E-72, with 72 inches between track centers. Hence, the EG-62 (G for gas) is a standard gasoline-powered Cletrac E with a track gauge of 62 inches.

In October 1944, Cletrac was purchased by Oliver Farm Equipment Company, which continued to market track tractors under the name

Eventually purchased by Oliver Farm Equipment Company, Cletrac derived its name as an acronym for the Cleveland Tractor Company. *Ralph W. Sanders*

Featuring a unique innovation marketed as Tru-Traction, Cletrac tractors were able to make turns while transmitting controlled amounts of power to each track, thereby maintaining full traction. *Ralph W. Sanders*

Oliver-Cletrac until 1965. The irony, however, is that in 1960, White Motor Corporation, which was descended from the White family that founded Cletrac, purchased the assets of Oliver, essentially bringing the Cletrac brand full circle.

Farmall Regular

As successful as farm tractors had become by the 1920s, there was still a desperate need on the farm for a tractor that could do it all. Every tractor to date was either both heavy and powerful for belt work and for pulling large implements, or it was light and maneuverable for cultivating row crops.

Years of production	1924–1932
Engine	International Harvester, 220.9 cubic inches (3.6 liters), 4 cylinders
Fuel	Gasoline
Horsepower	20.05 belt (tested), 13.27 drawbar (tested)
Rated RPM	1,200
Transmission	3 speeds forward, 1 reverse
Weight	4,100 pounds
Original price	Unknown

Recognizing the demand for a machine that could fill both roles, International Harvester began working on a "motor cultivator" as early as 1915. Unfortunately, it didn't work well and proved too expensive to build.

By 1924, however, the efforts of International Harvester engineers paid off in the form of a new tractor appropriately called the Farmall Regular. As the new Farmall name implied, it could do *all* that a farmer wanted it to do, from powering a thresher or sheller with its 20-horsepower belt pulley to pulling a plow with 13 horsepower at the drawbar. It even had a PTO to power a binder or sickle mower. Unlike other tractors of its time, the Farmall Regular also featured wide rear-wheel spacing, high ground clearance, and a narrow front end, which allowed it to straddle two rows while the front

Prior to 1924, tractors were primarily designed for either plowing or powering implements via a belt pulley. The Farmall Regular changed all that by becoming the first all-purpose tractor. Designed to also perform cultivation, the Farmall Regular additionally established the "tricycle" configuration. *Randy Leffingwell*

wheels ran between rows. This new configuration, which came to be known as the tricycle or row-crop design, gave farmers the ability to cultivate two rows of crops in a single pass with a machine that could perform a wide variety of other jobs on the farm.

Needless to say, the Farmall changed not only the future of International Harvester, but the future of the entire tractor industry worldwide.

Allis-Chalmers U

By itself, the Allis-Chalmers Model U was a very capable tillage tractor for fieldwork on a growing number of mechanized farms. Initially equipped with a 284-cubic-inch Continental engine, it developed 35 horsepower at 1,200 rpm. The same tractor was sold as the United by the United Tractor and Equipment Cooperative of Chicago, Illinois. That company had contracted Allis-Chalmers for a new tractor to replace the Fordson, which became unavailable to US and Canadian dealers when Ford stopped North American production.

The U/United went through two more engine changes, though, before it was discontinued in 1953 after a 24-year run. Beginning in 1932, at serial number 7405, the U was fitted with a UM four-cylinder, 300-cubic-inch engine, which was an Allis-Chalmers–designed engine built by Waukesha. Three years later, in 1936 at serial number 12001, the U was equipped with a 318-cubic-inch UM engine built by Allis-Chalmers.

The most noteworthy thing about the U, though, isn't the fact that it used three different engines during production. Instead, it is famous to this day for being the first tractor in the industry to use rubber tires. To test the idea, a 1932 Model U was fitted with truck tires on the front, while the rear tires were Firestone 12x48 tires from a Ford Tri-Motor airplane. The idea—which was a joint experiment by Harry Merritt, general manager of Allis-Chalmers from 1926 to 1941, and the Firestone Company—was a success.

However, while employees at Allis-Chalmers quickly discovered that tread-type rubber tires were just as effective as lugs, in addition to saving

Years of production	1929–1952
Engine	Continental, 284 cubic inches (4.6 liters), 4 cylinders
	Waukesha UM, 300 cubic inches (4.6 liters), 4 cylinders (SN 7405)
	Allis-Chalmers UM, 318 cubic inches (5.2 liters), 4 cylinders (SN12001)
Fuel	Gasoline
Horsepower	35.04 belt (tested), 30.37 drawbar (tested)
Rated RPM	1,200
Transmission	3 speeds* with steel wheels, 4 speeds with rubber tires
Weight	4,600 pounds
Original price	Unknown

* The number of gears was reduced by removing a shifter fork from the regular transmission.

Introduced in 1929, the Allis-Chalmers U featured steel wheels just like all the other tractors of its day. However, that all changed in 1932 when the U became the first tractor in the industry to feature rubber tires. *Ralph W. Sanders*

fuel and improving comfort, many farmers were still skeptical. To win them over, Allis-Chalmers initiated one of the most effective advertising campaigns in history. They hired Albert Schroeder, a Wisconsin farmer who purchased the first Model U with rubber tires, to travel the country giving talks and demonstrations. They also hired Barney Oldfield, one of the most famous racecar drivers at the time, to perform in tractor races, featuring Model U tractors on rubber, at many county and state fairs. Within a matter of years, every tractor sold in America was equipped with rubber tires.

Minneapolis-Moline UDLX

Sometimes it doesn't pay to be ahead of your time. That was certainly the case when Minneapolis-Moline introduced the UDLX, or U Deluxe, in 1938. Also called the Comfortractor, the unit was a standard U with an enclosed engine, full front and rear fenders, and an all-steel cab with a large door at the back for easy access. Even though other tractor brands were beginning to feature streamlined designs, nothing to this point quite matched the Comfortractor. Inside the roomy cab, prospective customers also found cushioned seats for two, a heater and ventilation, windshield wipers, a radio,

Years of production	1938–1941
Engine	Minneapolis-Moline, 283 cubic inches (4.6 liters), 4 cylinders
Fuel	Gasoline
Horsepower	6 PTO (claimed), 27 drawbar (claimed)
Rated RPM	1,275
Transmission	5 speeds
Weight	6,400 pounds
Original price	$2,155

The Minneapolis-Moline UDLX, or Comfortractor, was ahead of its time in every way possible. For starters, it had an all-steel, enclosed cab with all the comforts of a heater, seats for two, a radio, and an accelerator pedal. It was also designed to run up to 40 miles per hour on the road, allowing it to serve as both a tractor and automobile. *Ralph W. Sanders*

and other automotive amenities, including an accelerator pedal on the floor. Its five-speed transmission also featured a road gear that offered a road speed of 40 miles per hour. Consequently, the UDLX was designed to serve as both a tractor and an automobile.

Unfortunately, the UDLX never became popular with farmer customers—perhaps, in part, because of a price of $2,155—and only 125 units were built by the time it was removed from the lineup in 1941, making it one of the most collectible models in the Minneapolis-Moline lineage.

Ford Ferguson 9N

Years of production	1939–1942
Engine	Ford, 119.7 cubic inches (2.0 liters), 4 cylinders
Fuel	Gasoline
Horsepower	23.56 belt (tested), 16.31 drawbar (tested)
Rated RPM	1,500
Transmission	3 speeds
Weight	2,140 to 3,375 pounds
Original price	$585

From the very start in 1939, the Ford Ferguson 9N represented a true revolution compared to other tractors on the market. Ford had been in the tractor business earlier with the Fordson models. Yet, nothing in the Fordson line ever matched the first Ford N Series model in terms of popularity or sales volume.

One of the first things the 9N had going for it was the talents of the Ford design department, which created the stylish sheet-metal hood and grille that not only enclosed the engine and radiator, but also provided a hint of art deco. While other brands were moving away from wide front axles to a tricycle configuration, the Ford design embraced the wide front axle along with a wide, low profile that was extremely stable yet very maneuverable. Mounting a cultivator on the rear of the tractor, instead of on a high-clearance tricycle design, would additionally permit a lower operator station that was ahead of the axle for a smoother ride.

Ford was ahead of most tractors, too, in that it had a complete electrical system as standard equipment. That included a starter, battery, generator, and a direct-drive distributor with an integral coil. Rubber tires and a PTO were also standard.

Most importantly, however, the 9N boasted all of the revolutionary features developed by Harry Ferguson, including a three-point hitch system that changed the entire tractor industry, and automatic depth control. The three-point hitch allowed implements to be easily attached and removed, while the draft control system automatically regulated the working depth of the implement, which particularly helped on steep slopes.

For power, the 9N featured a 119.7-cubic-inch displacement, four-cylinder engine that was actually based on one half of the 239-cubic-inch Ford/ Mercury V-8 truck engine. Consequently, it used the same pistons, rings, valves, connecting rods, and bearings. The parts' commonality, combined with the benefits of mass production, which was actually a Henry Ford innovation, helped make the Ford Ferguson 9N one of the most affordable tractors on the market.

The Ford 9N, or more appropriately, the Ford Ferguson 9N, became one of the greatest superstars of its time when it combined the three-point hitch developed by Harry Ferguson and the tractor design team of Ford with the assembly line efficiency developed by Henry Ford. The handshake agreement that created the 9N became a true landmark in tractor history. *Ralph W. Sanders*

At a launch price of $585, it was priced much lower than tractors being offered by any of the competitors. Henry Ford's challenge to the engineering department was to build a tractor that would cost no more than the combined cost of a team of horses or mules, the harness, and the 10 acres of land required to produce enough feed for the team. At an estimated cost of $590 for the latter, the engineering team came out with $5 left over.

Minneapolis-Moline U-LPG

Years of production	1938–1959 (all U series models)
Engine	Minneapolis-Moline, 283.7 cubic inches (4.6 liters), 4 cylinders
Fuel	Gasoline, propane, and diesel
Horsepower	36 PTO (claimed), 27 drawbar (claimed)
Rated RPM	Unknown
Transmission	6 speeds forward, 1 reverse
Weight	14,500 pounds
Original price	$2,700

With much of their market in the Great Plains, where oilfields were prominent and liquefied petroleum gas (LPG) was cheap and plentiful, Minneapolis-Moline became a pioneer in the use of LPG as a fuel and began offering LPG-powered engines as early as 1941. By 1949, Minneapolis-Moline was the first to have an LPG-powered tractor in the Nebraska Tractor Tests when it submitted a U-LPG for testing. As a result of its success with propane and gasoline, Minneapolis-Moline discontinued the production of its low-compression, distillate-fueled, and kerosene-powered engines in 1954. A relative latecomer to the diesel engine market, Minneapolis-Moline finally introduced a diesel-powered U in 1952.

Equipped with a new hood and grille, the U was designed and styled to match a wide variety of needs, based on the model variation. The UTU, for example, was a row-crop model, which featured a narrow front. The UTS featured a standard front axle, the UB featured an improved operator platform, and the UTC came with dropped rear

Although Minneapolis-Moline was very successful with it's gasoline and distillate-powered tractors (left), the company began offering LPG-powered engines as early as 1941 and in 1949 became the first to have an LPG tractor tested at the Nebraska Tractor Tests when a U-LPG was submitted. *Ralph W. Sanders*

Styled much like the Universal Model Z, the Universal and standard Model U tractors were three/four-plow tractors with 42.7 engine horsepower developed by a 283-cubic-inch engine. Although gasoline and LPG were the most popular fuels (distillate-fueled engines were discontinued in 1954), a diesel engine option was added in 1952. *Ralph W. Sanders*

final drives and an arched front axle for the extra clearance cane producers needed. The U Series even included the famed UDLX, which featured an enclosed cab and automotive features.

Cockshutt 30/Co-op E-3

Although there is still some debate as to which company introduced the first independent or live PTO, Cockshutt was the first to receive the credit when the company tested the Model 30 at the Nebraska Tractor Test Laboratory in 1946. The greatest benefit of this new feature was the fact that the PTO did not stop operating when the operator pushed in the clutch to stop ground travel. An important advancement in tractor technology, the

With the development of the Model 30, Cockshutt literally went from a buyer to a seller. All their earlier tractors had been rebadged models built by other companies. However, with the Model 30, Cockshutt not only had its own tractor, but became a supplier of the Co-op E-3 (shown) to The National Farm Machinery Cooperative. *Ralph W. Sanders*

Years of production	1946–1957
Engine	Buda, 153 cubic inches (2.5 liters), 4 cylinders
Fuel	Gasoline, diesel, propane
Horsepower	30.28 PTO (tested), 27.25 drawbar (tested)
Rated RPM	1,650
Transmission	4 speeds
Weight	3,620 to 5,528 pounds
Original price	$2,500

independent PTO was particularly valuable when operating a loader powered by a PTO-driven pump or when pulling a PTO-driven implement, such as a sickle bar mower. No longer did farmers find themselves plugging a machine when the PTO drive stopped the moment they stopped the tractor. So it's little wonder other manufacturers found a way to incorporate the feature into the powertrain as soon as possible. As an example, the Oliver 88 is credited with being the first mass-produced US-built tractor with an independent PTO, and Minneapolis-Moline is said to have developed a tractor with a live PTO that was never marketed. Regardless of the source, a live PTO soon became a standard feature on farm tractors.

The Cockshutt 30 was historic in other ways, as well. After marketing tractors built by other companies, including Allis-Chalmers, Hart-Parr, and

Oliver, Cockshutt began developing its own tractors in the 1940s. The Model 30 became one of the first when it was introduced in 1946. It was initially available as a gasoline-powered model, but was available with a diesel engine in 1949 and with a propane-powered engine shortly thereafter.

The Cockshutt 30 was also marketed under other names, particularly in the United States. The National Farm Machinery Cooperative, for instance, began marketing the Cockshutt 30 as the orange-painted Co-op E-3. The Cockshutt 30 was also imported and sold by the Gamble-Skogmo Company, based in Minneapolis, Minnesota, as the Farmcrest 30. Most of those, which were painted red, were sold in the Upper Midwest.

The Cockshutt 30 made history in another way in that it was the first tractor Cockshutt actually designed and built themselves. Prior to the 30, the company had always marketed tractors built by other companies. *Ralph W. Sanders*

Although the same tractor was marketed as the Co-op E-3, the Cockshutt 30 received the credit with Nebraska Tractor Test Laboratory for being the first tractor in the industry to feature a live PTO. *Ralph W. Sanders*

A Departure in Design

Ever since the first tractor was introduced and marketed, tractor manufacturers have been looking for ways to improve their product. More often, though, it's meant coming up with a new design that makes their product unique in the industry, or at least a step ahead of their competitors.

Sometimes it has meant incorporating or designing an engine that ran on gasoline instead of kerosene or distillate—or even steam. Other times, it was building a small tractor to meet the needs of truck farmers and small operators who couldn't afford or use the bigger tractors that dominated the market.

Just like the tractors that were first in their field, some design modifications were simply ahead of their time. The Bates Steel Mule, with its half-track design, which still used front wheels to steer, comes to mind as one example. Today's tractor market offers articulated tractors with tracks in place of all four wheels, as well as conventional tractors with tires on the front and rubber tracks on the back.

The key to survival was recognizing a need and developing a product to fill it. It didn't help, though, that the Great Depression of the 1930s occurred right when the tractor industry was in the midst of its growth. If that wasn't enough, World War II brought tractor manufacturing to a near standstill as many farm equipment manufacturers shifted their resources to building military equipment and war supplies.

In the following pages, you'll find just a few examples of design modifications that happened at the right time in the right place—even if some of the manufacturers that introduced them didn't survive.

Andrew Morland

Ralph W. Sanders

Years of production	1910–1914
Engine	Wallis, 1,062 cubic inches (17.4 liters), 4 cylinders
Fuel	Gasoline
Horsepower	50 belt (claimed), 30 drawbar (claimed)
Rated RPM	650
Transmission	3 speeds forward, 1 reverse
Weight	16,000 pounds
Original price	Unknown

Wallis Bear

Built by the J. I. Case Plow Works, the Wallis Bear evolved from the design of a heavy tractor developed by Wallis in 1902. The irony is that Henry M. Wallis Sr. was the son-in-law of Jerome I. Case, founder of J. I. Case Plow Works, hence the manufacturing connection. Even though Wallis had taken over as president of Case after his father-in-law's death, he still had aspirations of building a gasoline tractor as an alternative to the steam traction engines for which Case had become famous. His experiments led to the development of two different designs of three-wheeled tractors and the formation of the Wallis Tractor Company in 1912. Individually known as the "Bear" and "Cub," the first models were advertised as "Fuel-Save Tractors."

Weighing in at nearly 16,000 pounds, the Bear was the largest. It introduced a number of features not found on any tractor. Most remarkable were the independent rear brakes and mechanical power steering via a pair of friction cones, which allowed the three-wheeled Bear to turn around within its own 12-foot wheelbase. In the meantime, an arrow on a rod coupled to the front wheel allowed the operator to see which way the wheel was pointed, since it couldn't be seen from the operator's seat.

Even though the drive sprockets that engaged the wheel were exposed, the transmission on the Wallis Bear was enclosed in a dustproof casing.
Ralph W. Sanders

The engine was unique, too, in that the four 6½x8-inch cylinders were cast in pairs and the engine was equipped with a speed governor. Plus, the transmission was enclosed in a dustproof casing. Even more prominent, though, was the large tubular radiator at the front of the tractor, which contained 460 feet of tubing. Topping off its massive dimensions were rear wheels that measured 7 feet in diameter.

Only a small number of Bear tractors were built, and it's believed that only one survives today.

The Wallis Bear was a big tractor in every sense of the word. Its large features included a 12-foot wheelbase and rear wheels that measured 7 feet in diameter. *Ralph W. Sanders*

Samson Sieve Grip

The Samson Sieve Grip got its name from the rear drive wheels, which were open between a series of cleats to, supposedly, provide better traction than solid wheels could in hilly terrain. Originally available in two models in 1914—a 6-12 model and a 10-25 model (shown)—the Sieve Grip also featured a wide, three-wheel design and a low profile that lent itself to orchard use. Unfortunately, it had a wide turning radius, despite having a single front wheel supported by a cast gooseneck. In fact, the operator couldn't see the front wheels due to the low position of the seat and the 1½ feet between the front wheel and the operator seat. Hence, an arrow was mounted to the top of the wheel to tell the operator which direction the front wheel was pointing.

Years of production	1914–1918
Engine	General Motors, 383 cubic inches (6.2 liters), 4 cylinders
Fuel	Kerosene
Horsepower	25 belt (claimed) 12 drawbar (claimed)
Rated RPM	700
Transmission	1 speed forward, 1 reverse
Weight	5,800 pounds
Original price	Unknown

With its wide fenders and low profile, it's little wonder the Samson Sieve Grip tractor found some of its earliest applications in the orchards around Stockton, California, where the tractor was first introduced. *Ralph W. Sanders*

On the plus side, however, Samson was one of the first tractor manufacturers to use enclosed gears to protect against wear from dirt and dust.

In 1918, Samson was purchased by General Motors in the automotive company's attempt to keep up with Ford. As a result, operations were moved to Janesville, Wisconsin, and consolidated

Since the tractor sat so low, preventing the operator from seeing the front wheels, Samson provided an arrow on the steering shaft to help with guidance. Of course, the GMC emblem denoted the fact that Samson was part of the GMC family in their attempt to compete with Ford. *Ralph W. Sanders*

The Sieve Grip name came from the fact that the steel wheels were open between the cleats, providing a sieve for dirt to move through as the wheels gripped the ground. *Ralph W. Sanders*

with the Janesville Machine Company operations, a manufacturer of farm implements that had also been acquired by General Motors. Following the move, the engine was changed to a four-cylinder General Motors model and a few structural alterations were made. Yet, General Motors continued to market tractors under the Samson name.

General Motors soon discovered, though, that the Sieve Grip tractors were too expensive to compete with Ford and eventually replaced them with a smaller, more affordable M. By 1922, high production costs, the agricultural depression of the early 1920s, and competition from Ford had taken enough toll that General Motors decided to abandon the tractor market. The company sold the remaining Samson inventory in 1923 and exited the agricultural enterprise altogether.

Allis-Chalmers 10-18

Like virtually every other farm equipment company in the early 1900s, Allis-Chalmers was looking for a way to enter the emerging tractor business. Although the company had already looked at a few options, including a "Motor-Driven Rotary Plow" and a "Tractor-Truck," a third option presented to the company was to enter into a joint agreement with Lyons, Knoll, and Hartsough in Minneapolis, Minnesota, to offer that company's Bull tractor.

However, the Bull tractor had only one drive wheel, which could easily stall in poor traction conditions. Ultimately, General Otto Falk, president of

Years of production	1914–1923
Engine	Allis-Chalmers, 303 cubic inches (5.0 liters), 2 cylinders
Fuel	Gasoline
Horsepower	18 belt (claimed), 10 drawbar (claimed)
Rated RPM	720
Transmission	1 speed forward, 1 reverse
Weight	4,800 pounds
Original price	$750 (1916)

Although its three-wheel design is similar to the Bull tractor, which Allis-Chalmers actually considered marketing, the 10-18 applied power to both wheels. It also had the single front wheel in line with the right rear wheel to better follow a furrow while pulling a plow. *Ralph W. Sanders*

Allis-Chalmers at the time, and the driving force behind the tractor push, chose to build and market an Allis-Chalmers design instead.

That model, of course, was the 10-18, which went into production in 1915. While it featured a three-wheel design, similar to the Bull tractor, the 10-18 not only applied power to both rear wheels, but went so far as to feature enclosed gears that ran in oil. The horizontally opposed two-cylinder engine ran at a relatively slow 720 rpm to produce 10 horsepower at the drawbar and 18 belt horsepower.

The other advantage the 10-18 offered was the fact that its single front wheel was in line with the right rear wheel, which meant both wheels could run in the plow furrow. Quoting from the original sales brochure, "The simple, durable, practical construction of the Allis-Chalmers Farm Tractor assures you unusual service. Operating costs are low for it is economical of fuel. The frame is a one-piece steel casting—no joints, no rivets, no weak spots. In plowing, it takes the place of six horses and will operate either two or three plows according to the condition of the soil plowed."

John Deere 62

In 1937 John Deere introduced a tractor that was dramatically different from the other models in the two-cylinder line. Billed as a one-plow utility tractor that could replace one or two horses, the 62 tested at 7.01 drawbar and 9.27 belt horsepower. More unusual than its low horsepower rating, though, was the fact that it featured a vertical two-cylinder engine rather than two horizontal cylinders. Unlike the other John Deere tractors of the time, it also featured a driveshaft and a foot clutch. Moreover, it wasn't built in Waterloo, Iowa, like the other models, but at the Wagon Works factory in Moline, Illinois.

Of particular interest to tractor collectors is the low production run of the 62. Only 78 units were built before it was redesigned, bumped up by four horsepower, and released as the L after only one year of production. The L continued in production until 1946 and is distinguished from the 62 by little more than the letters "JD" that appear below the radiator on the latter.

A few years later, in 1941, John Deere released the slightly larger Model LA. Not only did it have a bigger engine, but it also featured a 540-rpm PTO in addition to a belt pulley. Other than the PTO and larger rear tires, though, it is pretty hard to distinguish the LA from the L. Both models were finally discontinued in 1946.

Years of production	1937
Engine	Hercules NXA, 56.5 cubic inches (0.9 liter), 2 cylinders
Fuel	Gasoline
Horsepower	9.27 belt (tested), 7.01 drawbar (tested)
Rated RPM	1,550
Transmission	3 speeds forward, 1 reverse
Weight	1,500 pounds
Original price	$450

The John Deere 62, which was soon replaced by the L, was the first John Deere tractor to feature a vertical two-cylinder engine. A bevel gear set between the engine and the transmission provided power to a belt pulley on the left side of the tractor. *Andrew Morland*

Bates Steel Mule

Years of production	1937
Engine	Erd, 341 cubic inches (5.6 liter), 2 cylinders
Fuel	Kerosene
Horsepower	20 belt (claimed), 12 drawbar (claimed)
Rated RPM	1,100
Transmission	3 speeds forward, 1 reverse
Weight	4,300 pounds
Original price	$450

Although it was first built by the Joliet Oil Tractor Company in 1913, the Bates Steel Mule was designed by a man named Albert Bates, who got his start building machinery for wire fencing. His first tractor weighed 5,600 pounds, cost $985, and pulled three 14-inch plows.

Unlike other tractors of the time, though, the Steel Mule was a semicrawler with a 15-inch rear crawler track and steel wheels on the front for steering. A few years later, Bates Tractor Company merged with Joliet Oil Tractor Company to form Bates Machine & Tractor Company. Over the next few years, the company built several half-track models, including the D 12-20

Although the Bates Steel Mule came in several sizes, the D 15-22 was the most popular. The half-track design was supposed to provide more traction, while the steel front wheels provided more conventional steering. Of course, the "steel mule" name was no coincidence, as most farmers were still using horses and mules at a time when tractor companies were trying to convert them to mechanization. *Ralph W. Sanders*

(shown on page 41). Rated at 12 horsepower on the drawbar and 20 at the belt pulley, it featured a four-cylinder Erd engine with a 4x6-inch bore and stroke.

Competition eventually forced the company to concentrate on fully tracked models, which they produced well into the 1930s.

Avery Ro-Trak

It's hard for any product to be all things to all people, but Avery certainly gave it a good try when the company introduced the Ro-Trak in 1938. Featuring a modern, streamlined design, the Ro-Trak was unique in that each front wheel was mounted below a vertical cylinder, which helped the front

The Avery Ro-Trak was a rather odd-looking tractor that was actually designed with good intentions. The idea was that by rotating each front wheel 90 degrees, the owner could switch from a tricycle configuration to a standard-tread configuration. The vertical cylinders, meanwhile, were designed to absorb the shock loads in the absence of a pivoting axle. *Ralph W. Sanders*

wheels run over bumps and contours in the absence of a pivoting front axle. The support columns were, in turn, attached to a set of swing arms. This permitted the front wheels to be moved out to each side of the front of the tractor for a standard-tread configuration, or pulled together in front of the tractor for a row-crop configuration. As a row-crop tractor, the front wheel tread was 16 inches, which permitted the mounting of a cultivator.

However, when moved to the outer or standard-tread configuration, the front wheel tread measured 56 inches, which could be matched by the rear-wheel tread. This allowed both the front and rear wheel to run in the furrow for plowing. Still, Avery stated in their advertising that most farmers would need to change the position of the front wheels only twice a year—once for plowing, seedbed preparation, and planting, and once for cultivating the crop. Once it was changed back to a wide spacing for corn picking, combining, and/or grain hauling, it would be left that way until the following spring. Moreover, it was claimed that one man could change from one tread width to the other in as little as 30 minutes, using only two wrenches and a jack.

Years of production	1938–1941
Engine	Hercules QXB-5, 212 cubic inches (3.4 liters), 6 cylinders
Fuel	Gasoline
Horsepower	30 (estimated)
Rated RPM	2,000
Transmission	3 speeds forward, 1 reverse
Weight	4,000 pounds
Original price	Unknown

Cletrac General GG

The Cletrac General GG didn't represent a departure in tractor design in general as much as it did a design departure for Cleveland Tractor Company, or Cletrac. Up to the point the GG was introduced in 1939, Cletrac had built nothing but track tractors. Moreover, the GG was not only the first rubber-tired row-crop tractor the company built, but the only one.

Classified as a one-plow tractor, the GG was available only with a single front wheel. The same tractor was reportedly marketed as the Wards "Twin-Row" by Montgomery Ward and as a Co-op tractor in Indiana. Both of those versions were painted red, and their serial numbers were intermingled with the General GG as they came off the production line.

However, in the later part of 1941, Cletrac began concentrating its production for the war effort, and the manufacturing rights for the model GG were sold to the B. F. Avery Company of Louisville, Kentucky, in 1942.

Years of production	1939–1942
Engine	Hercules IXA2, 113 cubic inches (1.9 liters), 4 cylinders
Fuel	Gasoline or kerosene
Horsepower	22 belt (claimed), 15 drawbar (claimed)
Rated RPM	1,400
Transmission	3 speeds forward, 1 reverse
Weight	2,105 pounds
Original price	$650

The Cletrac General GG was the only wheeled tractor ever built by the Cleveland Tractor Company. Ironically, the GG was marketed under other brand names and went on to become part of the B. F. Avery lineup, as well as the Minneapolis-Moline tractor line. *Ralph W. Sanders*

After acquiring the model, Avery continued to use the GG letter designation for a short time, before changing it to the Model A. A few years later, in 1946, Avery added a wide-front version of the tractor known as the Model V. The small tractor took its final turn in 1951 when B. F. Avery was acquired by Minneapolis-Moline.

Farmall A

The Farmall A and its counterpart, the B, were designed and built in response to the Allis-Chalmers B, which had just been introduced. International Harvester needed a unit that could compete in the small tractor market. Even though its design cost a fee to Allis-Chalmers for partial patent

The Farmall A was designed in large part to compete with Allis-Chalmers in the small "truck garden" market. However, International Harvester put a twist on their design with their Cultivision chassis. In effect, the operator seat and controls were offset to the right, giving the driver a clear view ahead. *Ralph W. Sanders*

infringement for their use of a torque tube, International Harvester came up with its own unique innovation with its Cultivision design. It was specifically targeted toward small farmers, especially those who needed a small tractor for cultivating crops.

Basically, Cultivision amounted to offsetting the engine and driveline from the centerline of the tractor. The operator seat was then mounted on the right axle, providing the operator with an unobstructed view down and ahead of the tractor. The controls were mounted to the left of the operator, while the pedals were directly ahead.

While the A featured a wide front axle and no left axle casing, the Farmall B was nearly identical except for the fact that it had a narrow front axle with either one or two tires. The left axle casing also extended out to the left drop housing, which increased the width to handle two rows, instead of

Years of production	1939–1947
Engine	International Harvester, 113 cubic inches (1.9 liters), 4 cylinders
Fuel	Gasoline or kerosene
Horsepower	18.34 belt (tested), 16.32 drawbar (tested)
Rated RPM	1,400
Transmission	4 speeds forward, 1 reverse
Weight	2,105 pounds
Original price	$750 (1947)

one. Yet another version, the Farmall AV, was similar to the A but was a high-clearance model designed for vegetables (V) and taller row crops.

Of course, the A also had the distinction of being one of the first wheeled tractors to feature the new design developed by industrial designer Raymond Loewy. First introduced on the International Harvester TD-18 crawler, the design not only featured a new grille and streamlined, rounded hood, but a new International Harvester logo, which would appear for years to come.

Allis-Chalmers G

With its unique rear-engine design, the Allis-Chalmers G was not only a radical departure for Allis-Chalmers, but it represented an entirely new concept for the tractor industry in 1948. Unlike all other tractors at the time, the G had its small four-cylinder engine along with the transmission mounted behind the rear axle. The obvious advantage of such a design was unrestricted view of the implements, which were mounted under the frame, behind the front wheels. To make sure nothing obstructed the operator's vision, Allis-Chalmers engineers even utilized a yoke-shaped steering wheel, much like that found on airplanes.

On the other hand, the G was never intended for the Midwestern and grain-producing states. Instead, it was designed for truck gardeners, nurseries, and specialty-crop applications, where the G proved to be ideal for planting, fertilizing, and cultivation of vegetables, seedlings, and berries.

Although the concept and early prototypes started out with a straight, one-piece mainframe, the final design consisted of an arched, twin-tube frame, which allowed more flexible implement design and attachment, not to mention the sought-after visibility.

However, while the G was developed by West Allis, Wisconsin, tractor engineering, it was actually built in Allis-Chalmers's Gadsden, Alabama, plant, which also built cotton pickers, mowers, and so on. Moreover, since Allis-Chalmers didn't build an engine small enough for the G, the company went to a Continental powerplant. Mounted behind the rear axle, it provided

Years of production	1948–1955
Engine	Continental, 62 cubic inches (1.0 liter), 4 cylinders
Fuel	Gasoline
Horsepower	10.33 belt (tested), 9.04 drawbar (tested)
Rated RPM	1,800
Transmission	3 speeds forward, 1 reverse
Weight	1,285 pounds
Original price	$760 (1949)

Even after more than 60 years of tractor history, the Allis-Chalmers G remains unique in its design. Because the engine and drivetrain are positioned behind the operator—and the implement is mounted under the frame—the driver was afforded an unrestricted view of the implement and field. Allis-Chalmers even developed a full line of implements to fit its unique tractor. *Andrew Morland*

extra weight on the rear wheels for added traction, especially when combined with a full fuel tank and the operator's weight.

Making the tractor even more appealing to truck farms was the fact that both front and rear wheels could be adjusted from 36 to 64 inches for row-crop adaptability. Plus, the special four-speed transmission offered an extra slow speed of 1.6 miles per hour, making it what some called a "hoe on wheels."

Time
FOR A
Change

I't's a recognized fact in nearly every industry that those who fail to keep up with changes in technology and customer desires will quickly fall behind. Nearly every type of business has its innovators, early adopters, and followers. Those facts have been true throughout much of tractor history, as well. While some changes came too early or proved to be failures, others caught on quickly and proved to be profitable for both the company and its customers.

The Farmall Regular, which was basically the first multipurpose, row-crop tractor in the industry, was a good example. As soon as it proved to be successful, other companies were all but forced to follow suit, including John Deere, which developed its GP in response to Farmall's success. The same has held true on tractor size and horsepower. When one company came out with a larger tractor, other manufacturers were almost forced to match it in order to keep up. However, there were times the opposite was true, like when Allis-Chalmers came out with the G for truck farmers. Other brands that were only known for higher-horsepower models were quick to follow, knowing that they, too, needed a smaller model to match the competition.

There were times, though, when a company didn't have the resources to change quick enough. You'll notice throughout tractor history that those were the times management decided to import a smaller model from Europe or Asia and put their own company's name on it—or have another American-based company build a tractor for them. That was certainly the case when Massey Ferguson initially needed a larger tractor to meet the growth in horsepower. For Massey Ferguson this meant buying tractors from Oliver and Minneapolis-Moline until it could develop its own high-horsepower model.

It wasn't always tractor size and horsepower that forced a change, though. In the mid- to late 1930s, it was style and color. It's been said that Allis-Chalmers switched from green paint to its infamous Persian orange in 1929 after Harry Merritt, manager of the tractor division at the time, saw a field of brilliant orange poppies while on a trip to California. Of course, that's just one story concerning the origin of the paint color. Whatever the reason, the bright orange paint quickly drew the attention of farmers and competitors alike. A farmer could tell from a mile away that his neighbor had bought a

Ralph W. Sanders

new Allis-Chalmers tractor, and that was advertising that money couldn't buy. Before long, every tractor manufacturer was looking at color as a selling point.

In the meantime, tractor manufacturers were restyling their tractors in response to the modern styling that was catching attention in the automotive world. Several companies, including John Deere and International Harvester, even hired professional design companies to help with their restyling efforts.

Today, tractor manufacturers continue to change designs and models to meet the needs of customers and the market. While change often revolves around horsepower, just as it did 100 years ago, many of the changes also involve the incorporation of new technology, guidance, and control. Tractors have also been forced to change in recent years to meet government regulations on emissions.

Perhaps the ancient Greek philosopher Heraclitus (535–475 BCE) said it best when he commented, "There is nothing permanent except change." That's certainly been true in the agricultural industry.

John Deere GP

Years of production	1928–1935
Engine	John Deere, 311 cubic inches (5.1 liters), 2 cylinders
Fuel	Kerosene
Horsepower	25.36 belt (tested), 18.86 drawbar (tested)
Rated RPM	900
Transmission	3 speeds forward, 1 reverse
Weight	3,600 pounds
Original price	$1,200

With the introduction of general-purpose row-crop tractors by some of John Deere's toughest competitors, including International Harvester, which had released the row-crop Regular a few months earlier, Deere couldn't ignore its need for a more versatile model. After two years of testing prototypes, the company released the GP in 1928.

The tractor was initially introduced as the C in 1927, but the name was changed due to the similarity to the existing D. Most of those models were eventually recalled and rebuilt. In the meantime, the company selected the name General Purpose, or GP, as a marketing ploy to imply that, like the Farmall Regular, it could do everything from plowing and belt work to cultivating. Like the extremely popular D, the GP featured a standard configuration, but with an arched front axle and drop gear housings on the rear axle to give it row-crop clearance. It was also smaller, with a 10/20 horsepower rating compared to 15/27 on the D. Finally, it was the first tractor

Faced with stiff competition from the Farmall Regular, which was designed as an all-purpose tractor, John Deere countered with the GP, or "general-purpose" tractor, in 1928. *Ralph W. Sanders*

to feature a motor-driven power lift for raising a mounted cultivator and planter.

The GP had a few shortcomings, though, which prompted Deere to introduce the GPWT, or Wide Tread tricycle design, in 1929. The most obvious features of the redesign were the 76-inch rear axle, tricycle front, and improved visibility, which was further enhanced by the 1932 version that had an engine hood that tapered in the

The GP was unique among John Deere tractors in that it was the only two-cylinder tractor to use the L-head valve configuration. *Ralph W. Sanders*

back. During its eight-year production run, various other versions of the GP were produced, including the GPO orchard version, the GP-P potato version, and crawler-tracked versions assembled by Lindeman Power Equipment Company in Yakima, Washington.

Minneapolis-Moline Z

Years of production	1937–1948
Engine	Minneapolis-Moline, 186 cubic inches (3.0 liters), 4 cylinders
Fuel	Gasoline or distillate
Horsepower	21.14 belt (tested), 23.39 drawbar (tested)
Rated RPM	1,500
Transmission	5 speeds forward, 1 reverse
Weight	3,650 pounds
Original price	$1,500 (1948)

Like many tractor manufacturers in the 1930s, Minneapolis-Moline was making the move to new styling and color to attract new customers and to keep pace, to a lesser extent, with the automotive industry. While other companies were going from gray or dark green to orange or red, Minneapolis-Moline chose Prairie Gold for the chassis and sheet metal on its new Z Series. In the meantime, the wheels were painted red.

Minneapolis-Moline had its own name for the styling—"Visionlined." Billed as "the last word in modern tractor design," this styling featured a sleek new contour with a narrow fuel tank and hood, which did help improve visibility, especially when cultivating. The Z also sported a new engine that, at 186 cubic inches (206 late models), was actually smaller than the engine used in its predecessor. Yet, it was more powerful. One of the engine's unique features was that the valves were mounted horizontally and controlled by long rocker arms. This was to make clearance adjustments much easier. In addition, the cylinder head could be removed and spacers added or removed to adjust the compression ratio, allowing the use of gasoline or distillate.

Two versions of the Z were initially available when the model was introduced in 1937. They included the ZTU with dual narrow front wheels and the ZTN with a single front wheel. A ZTS standard model with fixed treads and full fenders followed a year later for the wheat belt. In the meantime, the row-crop Z models featured crown fenders from 1938 through 1948, at which point the ZA model appeared with another round of new styling, shell fenders, and five more horsepower from a new 206-cubic-inch engine. To this point, all the Z models had the steering wheel offset slightly to the left of the seat so the steering shaft could run along the left side of the engine.

Introduced in 1937, the Minneapolis-Moline Z was one of the first Minneapolis-Moline tractors to feature the Prairie Gold color and the new "Visionlined" styling. All total, the Z went through two more revisions before being discontinued in 1948. *Ralph W. Sanders*

A final revision of the Z was marketed from 1953 through 1956 as the ZB. In addition to more horsepower, one of its key features was a higher operator platform and a centered steering wheel, which allowed the tractor to accept a new two-row tractor-mounted corn picker and the operator to better see and operate the attachment.

Even though the 186-cubic-inch engine in the Z was smaller than the one used in its predecessor, it was actually more powerful, providing just over 23 horsepower at the drawbar. *Ralph W. Sanders*

Farmall M

Years of production	1939–1954
Engine	International Harvester, 247.7 cubic inches (4.1 liters), 4 cylinders
Fuel	Gasoline or distillate
Horsepower	36.66 PTO (tested), 33.1 drawbar (tested)
Rated RPM	1,450
Transmission	5 speeds forward, 1 reverse
Weight	4,858 to 6,770 pounds
Original price	$1,200 (1939)

In the late 1930s, International Harvester became one more of the tractor manufacturers that began adding style to their products. The entire issue started in 1935 when Oliver Hart-Parr introduced the Oliver 70 with its sleek design that included louvered side panels and an enclosed grille. Within two years, other sleek designs were showing up, including the heavily chromed Graham-Bradley.

Not to be outdone, International Harvester brought in famous industrial designer Raymond Loewy. As the designer who would later develop the famous Studebaker car designs, Loewy was commissioned to redesign the

Thanks to its new design, which was contributed in large part by industrial designer Raymond Loewy, the Farmall M became one of International Harvester's best-selling models. Initially available with a gasoline or distillate engine, LPG and diesel (shown) models eventually followed. *Ralph W. Sanders*

entire line of International Harvester products, including the Farmall tractors, as well as the company logo.

The first fruit of his labors was the new TD-18 crawler tractor, introduced in 1938. Within a matter of months a new line of wheeled tractors followed, starting with the A. Right behind it were the H, which replaced the F-20, and the M, which replaced the F-30. Like the other redesigned models, the M featured bright red paint, a contoured sheet-metal hood that enclosed the fuel tank and steering bolster, and a new grille that enclosed the radiator. The new tractors also boasted new safety features, like standard-equipment fenders and controls that were easier to reach.

John Deere H (Styled)

One of the biggest changes affecting John Deere tractors occurred in 1938 when Deere introduced its newly streamlined and styled tractor models. The transformation actually began in 1937 when Deere hired a Madison Avenue design firm, known in the industrial design world as Henry Dreyfuss Associates, to help style its two most popular models, the A and B.

Born in New York, Henry Dreyfuss had apprenticed as a designer of theater costumes, scenery, and sets before establishing himself as an industrial designer. Still, tractors were about as foreign to the New York designer as Madison Avenue was to most farmers.

That didn't deter Dreyfuss, however, from doing what he called a "clean-up" on the John Deere tractor design. In his initial efforts alone, he produced a redesigned hood that surrounded the steering gear and developed a grille that enclosed the radiator. Although the new styling was eventually extended to the M and the D, the first Deere tractor to feature the new styling from its inception was the Model H, introduced in 1939.

Billed as a two-row tricycle model, it was built from 1939 to 1947 and was produced in numbers exceeding 60,000. Unlike the other letter models, though, the H had a few peculiarities. One was the fact that power was taken off the camshaft, rather than the crankshaft. This meant that the belt pulley

Years of production	1939–1947
Engine	John Deere, 99.7 cubic inches (1.6 liters), 2 cylinders
Fuel	Kerosene
Horsepower	14.84 belt (tested), 12.48 drawbar (tested)
Rated RPM	1,400
Transmission	3 speeds forward, 1 reverse
Weight	2,080 to 3,035 pounds
Original price	$650 (1947)

Introduced in 1939, the Model H became the first John Deere model to feature the new styling developed by New York design firm Henry Dreyfuss Associates. The new styling was later implemented on the rest of the John Deere line, including the restyled A and B. *Ralph W. Sanders*

ran in the opposite direction from that of crankshaft-driven belt pulleys. It also meant that the bull gears were not required, which allowed the brakes to be applied directly to the axles.

Case S/SC

Years of production	1941–1952
Engine	J. I. Case, 153.9 cubic inches (2.5 liters), 4 cylinders
	J. I. Case, 165.1 cubic inches (2.7 liters), 4 cylinders (1953 on)
Fuel	Gasoline
Horsepower	31.71 belt (tested), 27.68 drawbar (tested)
Rated RPM	1,550, 1,660 (1953 on)
Transmission	4 speeds forward, 1 reverse
Weight	4,200 pounds
Original price	$1,700 (1955)

J. I. Case was in need of a couple of things when it introduced its new S and SC models in 1941. It needed an update of the R Series, which had already been modernized with new sheet metal and the Flambeau Red paint. Yet, it still had a Waukesha engine, rather than a Case-built engine. However, the company also needed a smaller version of the DC, which was rated as a three-plow tractor, to compete with the John Deere B and Farmall H. By making the S and SC as standard-tread and narrow-front row-crop tractors, respectively, Case figured to have the market covered on two fronts. The hot

While farm mechanization was allowing some farms to grow larger, farm equipment manufacturers were also seeing the demand for smaller tractors to fill the needs of small-acreage operations. Case was one of the latecomers to that market with the S/SC when it saw the need for something smaller than their DC model. *Ralph W. Sanders*

red-orange paint and new styling were enough to catch anyone's attention, but the two-plow rating was just what many farmers were looking for in a row-crop tractor, particularly those who used a mounted cultivator. That might explain, too, why the SC sold nearly six times as many tractors as the S did during the war years.

The SC could also be ordered with a dual-wheel tricycle front or a single front wheel. An additional SO was offered as an orchard model. All versions featured a new short-stroke, high-speed Case engine that provided reasonable power for the weight. However, Case engineers decided to give it even more power in 1953, when they increased the bore by 0.125 inches, adding another nine horsepower on the belt.

Farmall 560

Up until the Farmall 560 was introduced in 1958, the International Harvester line of tractors hadn't really changed that much in appearance. Granted, the letter series was replaced by a new number series in 1954, but the Farmall 300 didn't really look that much different than the H it replaced, and the 400 and 450 weren't that much different in overall appearance from an M, other than the updates and improvements.

However, that all changed with the 560 and the rest of the line of big tractors, as there was very little similarity between the 560 and the 450. Not only did the tractors feature a fresh new design with square lines, a new paint scheme, and a modern grille, but several of the models, including the 560, boasted a new six-cylinder engine. At the time it was introduced, the 560 claimed to be the world's most powerful row-crop tractor. It was also available with a gasoline, LPG, or diesel engine. What's more, the D-282 engine in the 560 diesel relied on glow plugs when starting as opposed to the switchover system that was dropped in 1958.

In addition to the modern design, the 560 and its stablemates also featured a wide seat with a backrest in place of the old pan seat and power steering. Optional features included the torque amplifier—which essentially

Years of production	1958–1963
Engine	International Harvester, 263 cubic inches (4.3 liters), 6 cylinders (gas/LPG) International Harvester, 281 cubic inches (4.6 liters), 6 cylinders (diesel)
Fuel	Gasoline, diesel, or LPG
Horsepower	63.03 PTO (tested), 53.12 drawbar (tested)
Rated RPM	1,800
Transmission	5 speeds forward, 1 reverse
Weight	6,563 to 9,460 pounds
Original price	$5,500 (1963)

doubled the number of speeds—a faster first gear and reverse, and International Harvester's exclusive Fast Hitch.

Unfortunately, the one thing that International Harvester did not change was the final drive system, which was carried over from the previous series. It simply wasn't strong enough to handle the increased power from the six-cylinder engine, which led to a number of drivetrain failures. International Harvester offered an entirely new rear end in mid-1959 and issued a massive replacement program. In an attempt to restore customer confidence, International Harvester also doubled the warranties and repaired tractors as fast as possible. Despite company cash flow problems at the time, they even provided loaner tractors to keep the customers going. Yet the damage had already been done in the minds of a number of customers. After additional

The Farmall 460 and 560 (shown), which were introduced in 1958, represented the first real design change for International Harvester tractors since the company introduced the styled letter series nearly 20 years earlier. *Ralph W. Sanders*

problems with the tractors involving clutch failure, steering problems, and more, a number of customers switched brands, allowing John Deere, which introduced its New Generation models approximately one year later, to pass International Harvester for the first time in history.

John Deere 4010

On August 30, 1959, the John Deere Company revealed one of the best-kept secrets the ag world has ever seen when it introduced its New Generation of four- and six-cylinder tractors to replace the two-cylinder tractors that had served the company since 1914.

The John Deere marketing and engineering teams had long realized that the horizontal two-cylinder engine had been developed about as far as it could go. The decision to develop a new line of vertical-cylinder-engine tractors was made in 1953. The project was so secret that engineers worked in an old grocery store known as the "Butcher Shop" in Waterloo, Iowa. To claim the New Generation tractors were "totally new" would be an understatement, as they bore little resemblance to earlier John Deere models. In fact, the company went so far as to have the New Generation tractors designed by Henry Dreyfuss and Associates, the same company that designed the styled letter-series tractors back in the late 1930s and early 1940s. Realizing that comfort was not only appreciated but required to alleviate the back pain caused by hours and days of riding over rough fields in a hard pan seat, John Deere employed

Years of production	1960–1963
Engine	John Deere, 302 cubic inches (4.9 liters), 6 cylinders (gas/LPG)
	John Deere, 380 cubic inches (6.2 liters), 6 cylinders (diesel)
Fuel	Gasoline, diesel, or LPG
Horsepower	84 PTO (tested), 71.93 drawbar (tested)
Rated RPM	2,200
Transmission	8 speeds forward, 3 reverse
Weight	7,100 to 9,595 pounds
Original price	$4,116 to $5,285

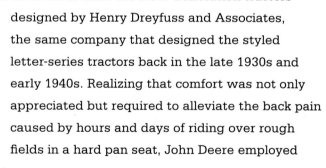

The 4010 was the largest of the conventional New Generation tractors introduced in 1960. In addition to being available with a gasoline, LPG, or diesel engine, it was available in several configurations, including row crop, high crop, standard tread, and industrial. *Ralph W. Sanders*

The premier of the John Deere New Generation tractors was one of the best-kept, if not the best, secrets in tractor history. The new line, which consisted of the 1010, 2010, 3010, 4010, and 8010, was introduced to 6,000 dealers and guests at Deere Day in Dallas, Texas, on August 30, 1960. It was the first introduction meeting in the farm equipment industry with all the company's dealers present at the same time. *Ralph W. Sanders*

the services of an orthopedic physician and back specialist to design a new seat for the roomy operator's station.

The initial line of new models consisted of the 1010, 2010, 3010, 4010, and 8010. While the two smallest models were built in Dubuque, Iowa, the other three models were assembled in Waterloo, Iowa. The 2010, 3010, and 4010 each

Since nearly 95 percent of the parts on the New Generation tractors were different from those used in the two-cylinder models they replaced, it also meant the factory had to be redesigned and upgraded. *Ralph W. Sanders*

featured a new Syncro-Range transmission that provided eight forward and three reverse gears. The synchronized transmission also allowed operators to change speeds on the move in each of four ranges and to shuttle-shift between forward and reverse. The PTO, meanwhile, offered speeds of either 540 or 1,000 rpm.

John Deere changed a lot more than the tractors when the company introduced the New Generation models. Since nearly 95 percent of the parts were different from those used in the two-cylinder models, it also meant the factory had to be redesigned and upgraded, thus setting the stage for efficient manufacturing and quicker adaptability for years to come.

Massey Ferguson 1150

Years of production	1970–1972
Engine	Perkins, 503.8 cubic inches (8.3 liters), 8 cylinders
Fuel	Diesel
Horsepower	135.60 PTO (tested), 125.63 drawbar (tested)
Rated RPM	2,200
Transmission	Multi-Power with 12 speeds forward, 4 reverse
Weight	11,000 pounds
Original price	$14,000

When the Massey Ferguson 1150 was introduced in 1968, it became the largest tractor in the company's line at the time, as well as the first Massey Ferguson tractor with a V-8 engine. It was also among the first big tractors actually built by Massey Ferguson. Prior to the introduction of the 1100 in 1964, the company had concentrated on small and mid-size tractors and had purchased its high-horsepower tractors from other manufacturers, including Oliver and Minneapolis-Moline. The 1100—the first of the 1100 Series—was the first high-horsepower tractor built in-house.

The Massey Ferguson 1150 and slightly smaller 1130 also reflected another change, led by customer demands. As rear-mounted implements became more popular, tractor manufacturers began to move the operator's platform forward on the chassis. As a result, the low-platform models were being relegated more and more to specialty markets. High-platform models, which gave the operator a better view of the field and implement, were making a comeback.

The 1100 was, consequently, one of the first Massy Ferguson models that positioned the operator at a vantage point that was higher up and ahead of the rear axle. The one thing that didn't change was the restyled design that featured squared hoods and four- or five-bar grilles. The new look had been

Massey Ferguson proved it was serious about horsepower when it introduced the 1100 Series in 1964. On its own, the MF1100, which launched the series, was the company's largest gas model, and the 1150 became the company's first V-8 diesel powered tractor. *Andrew Morland*

introduced with the smaller Massey Ferguson 135, 150, and 165 tractors introduced the same year.

Thanks to its 503.8-cubic-inch engine and standard equipment turbocharger, the 1150 certainly wasn't short on power at 135.60 PTO horsepower. The 1150 wasn't short on amenities, either. Hydraulic power brakes and power steering were standard, as was a 12x4 Multi-Power transmission. In the meantime, a pressure-control hydraulic system provided weight transfer from towed implements for improved field traction and performance. A category three three-point hitch was, of course, optional.

The pressure-control hitch, introduced on the 1100 Series, allowed the operator to transfer implement weight to the tractor's rear wheels for improved traction. *Andrew Morland*

Years of production	1964–1972
Engine	Allis-Chalmers, 265 cubic inches (4.3 liters), 6 cylinders
Fuel	Gasoline or diesel
Horsepower	77.20 PTO (tested), 65.33 drawbar (tested) (diesel)
Rated RPM	2,200
Transmission	8 speeds forward, 2 reverse
Weight	7,660 pounds
Original price	$7,175 (gas), $8,850 (diesel)

Allis-Chalmers 190

Thanks to the modern styling of the Allis-Chalmers D21, introduced in 1963, there was already a demand for new styling on the rest of the tractor line. However, the design team that created the D21 went far beyond the existing ideas when it proposed the replacement for the D19. What it ended up with was a functional combination of the D19 and D21 and styling that was unlike either one.

Often referred to as "the 190," rather than the spelled-out "One-Ninety," as the decal reads, the new model's name was an attempt to retain the

The Allis-Chalmers 190 brought all-new styling to the company's lineup of high-horsepower tractors. While many of the features introduced on the D21 were carried over to the new series, the 190 had several new features of its own. The 190XT versions soon followed, providing horsepower levels that were up to 20 percent higher. *Andrew Morland*

D19's number 19 and build on its popularity. In the meantime, the One-Ninety borrowed plenty of features from the D21, including a large, flat operator's platform, full fenders, a tilt steering wheel, and an adjustable seat mounted on an inclined track. It also had the fuel tank mounted behind the operator's seat.

The One-Ninety also featured the same low-pressure, high-volume hydraulic system from the D21, an engine-driven pump that supplied the power steering and Traction-Booster system, for a total of 12 gallons per minute at 2,000 pounds per square inch (psi) to the remote valves. One thing that was totally new, however, was the direct-injection diesel engine that was actually the first application of the 2500–2800 Series Allis-Chalmers diesel engines in a farm equipment application.

While the diesel-powered model came with a 301-cubic-inch engine, the gas version featured a Model G2500, 265-cubic-inch, six-cylinder powerplant. Both versions were available in several different configurations, including a single-wheel front axle, narrow dual-wheel front axle, and high-clearance adjustable front axle.

AGCO-Allis 9690

The 9600 Series represented a big change for orange tractors when it was introduced in 1993. Granted, it consisted of a totally new line of 135- to 195-horsepower models with a new 18-speed powershift transmission. More importantly, though, it represented the first use of the "AGCO" name on the side of a high-horsepower tractor since AGCO Corporation purchased the Deutz-Allis brand from the German company Klöckner-Humboldt-Deutz in 1990. Just five years before that, Klöckner-Humboldt-Deutz had acquired the remnants of Allis-Chalmers, ending a tractor legacy that dated back to 1914.

Consequently, the first thing AGCO did after bringing the brand back to American ownership was change the color of the existing 9100 Series from Deutz green back to Allis-Chalmers orange. However, little else had

Years of production	1993–1996
Engine	Deutz, 584 cubic inches (9.6 liters), 6 cylinders, air-cooled
Fuel	Diesel
Horsepower	195.6 PTO (tested), 182.8 drawbar (tested)
Rated RPM	2,300
Transmission	18 speeds forward, 9 reverse
Weight	15,200 pounds
Original price	$107,000 (1995)

The AGCO-Allis 9600 Series represented the first new tractor design for the newly formed AGCO Corporation. More importantly to Allis-Chalmers enthusiasts, it represented a return to the Allis-Chalmers orange color. Available with a liquid-cooled or air-cooled engine, the 9600 Series was also the first tractor with an armrest-mounted control console. *Ralph W. Sanders*

changed until the 9600 was introduced. In addition to the AGCO-Allis logo that replaced the Deutz-Allis decal, the 9600 featured new rounded sheet metal and a redesigned cab with more comfort and a unique, new seat-mounted control console that moved with the seat. The tractor also featured the new electronically controlled 18x9 powershift transmission that could be shifted one gear at a time by simply bumping the shift lever or shifted sequentially through the gears as the operator held the lever in upshift or downshift position.

One thing that didn't change was the Deutz air-cooled engine, which was purchased from the previous owner. The 135-horsepower 9630 and 155-horsepower Model 9650 used a 374-cubic-inch model, while the 175-horsepower Model 9670 and 195-horsepower Model 9690 used a 584-cubic-inch Deutz diesel. The 9690 was only available with front-wheel assist, while the other three models were available with two-wheel drive or all-wheel drive.

Love it or hate it, the air-cooled engine was soon joined by a liquid-cooled option just a year later when AGCO offered the 9605 Series with a Detroit Diesel Series 40 engine. The air-cooled models were eventually phased out as emissions standards became more stringent and the more popular liquid-cooled engines took over the market.

For more than 100 years, one thing has not changed. That is the need for more horsepower and bigger tractors. In fact, it's often hard to tell which came first; bigger implements that called for more powerful tractors, or tractors with more horsepower, which could pull larger equipment. Either way, farmers were finding that they needed to cover more acres in less time to remain profitable. That equation remains the same today.

Still, bigger tractors and more horsepower bring with them a new sense of awe. Just as today's rigid-frame tractors are now topping 300 engine horsepower, it was no less impressive when the muscle tractors of the 1960s were pushing past 100 horsepower.

First it was six-cylinder tractors in place of the traditional four-cylinder models. Then came the move to diesel engines and turbochargers. If that wasn't enough, one could always squeeze in a larger engine—say, one with 426 cubic inches.

Either way, these were some of the power superstars we won't soon forget.

Ralph W. Sanders

Andrew Morland

Years of production	1935–1948
Engine	Oliver, 201 cubic inches (3.3 liters), 6 cylinders
Fuel	Gasoline or distillate
Horsepower	30.37 belt (claimed), 22.7 drawbar (claimed)
Rated RPM	1,500
Transmission	4 speeds forward, 1 reverse (6 speeds after 1937*)
Weight	4,400 pounds
Original price	Unknown

* The top two gears were locked out on models purchased with steel wheels.

Hart-Parr Oliver 70 Row-Crop

When it was introduced in 1935, the Hart-Parr Oliver 70 Row-Crop tractor was certainly the right tractor at the right time for the market. Featuring a streamlined grille and side panels that enclosed the engine, it was more stylish than most other tractors on the market. Making it even more appealing was the fact that lights and an electric starter were optional.

Even more important, it was one of the first tractors on the market to feature a smooth-running six-cylinder engine that developed 28.4 horsepower on gasoline and 26.7 horsepower on kerosene. That fact alone contributed to

The Hart-Parr Oliver 70 was ahead of its time with more-than-sleek styling. It was also one of the first tractors on the market to utilize a six-cylinder engine and a four-speed transmission. *Ralph W. Sanders*

Oliver being the first tractor builder in the industry to truly exploit the extra power and performance that gasoline offered.

Coupled to the reliable engine was a four-speed transmission that offered an amazing top speed (at the time) of 6 miles per hour. Meanwhile, a mechanical power lift was capable of handling a two-row cultivator or other front-mounted implement.

For the growing number of medium-sized farms that survived and expanded following the 1930s, the 70 Row-Crop was the ideal model for those ready to embrace modern farm mechanization. By 1937, however, the Hart-Parr label had disappeared and the 70 carried only the Oliver name, along with a newly restyled grille and hood that was even more advanced than the competition. While the 201-cubic-inch engine continued, the tractor now featured a new six-speed transmission with speeds ranging from 2.5 to 13.4 miles per hour. Standard, orchard, and industrial models were also offered.

John Deere R

With a displacement of 416 cubic inches, the engine in the John Deere R was big in anyone's book. However, it's even more impressive when one considers that the R still had one of John Deere's infamous two-cylinder engines connected to the powertrain. Granted, the Model R was the first diesel-powered tractor that John Deere offered; with a bore and stroke of 5¾x8 inches, the pistons were said to be the size of coffee cans. Hence, it couldn't be started by hand crank or even by electric starter. Instead, it used a 24.6-cubic-inch, two-cylinder gasoline starter engine, or pony engine, to get it fired up.

Designed for pulling big tillage equipment and large-moldboard plows, the R was only available as a standard-tread, or wheatland, version. When the R was introduced in 1949, it was thought that it would replace the D, which had been in production since 1923. However, due to the popularity of the D and a last-minute boost in sales, John Deere continued the D until 1953.

Years of production	1949–1954
Engine	John Deere, 415 cubic inches (6.8 liters), 2 cylinders
Fuel	Diesel
Horsepower	48.58 belt (tested), 43.15 drawbar (tested)
Rated RPM	1,000
Transmission	5 speeds forward, 1 reverse
Weight	7,400 to 7,600 pounds
Original price	$3,650 (1954)

With pistons described as being the size of coffee cans, the John Deere Model R became the first diesel-powered tractor in the John Deere fleet. Designed for pulling large tillage equipment, it was also the most powerful. *Ralph W. Sanders*

In 1955, John Deere increased the displacement even more, to 472 cubic inches, increased the number of forward gears from five to six, and changed the designation from R to 80. Two years later, in 1957, the tractor received another 10 percent increase in power and came out as the 820, followed by the 830 a year later.

Oliver Super 99

Introduced in 1957 with a choice of three different engines, the Oliver Super 99 was, at the time, the largest two-wheel-drive tractor in the industry. When equipped with the optional General Motors (Detroit Diesel) 3-71 two-stroke, three-cylinder supercharged diesel engine, the Super 99 cranked out more than 78 horsepower. Even though the engine featured a small 213 cubic inches of displacement, its three power strokes per revolution provided enough muscle to take the Super 99 into the five-six plow class. Key to its horsepower rating was its 17:1 compression ratio and a Roots-type blower that helped purge exhaust gases and supercharge the fresh mixture.

The other option was the six-cylinder, 302-cubic-inch gas or diesel engine that was previously used in the 99. That engine cranked out just over 62 horsepower. The difference was that the Super 99 had a new six-speed transmission that brought with it improved field performance. First built in South Bend, Indiana, the Super 99 was later transferred to the Charles City, Iowa, plant in 1958. Available with a cab as an option, the Super 99 was built in both agricultural and industrial models.

Years of production	1957–1958
Engine	Waukesha-Oliver, 302 cubic inches (4.9 liters), 6 cylinders
	General Motors, 213 cubic inches (3.5 liters), 3 cylinders with blower
Fuel	Gasoline or diesel
Horsepower	78.74 belt (tested), 73.31 drawbar (tested) (Oliver)
	62.39 belt (tested), 58.27 drawbar (tested) (General Motors)
Rated RPM	1,675
Transmission	6 speeds forward, 2 reverse
Weight	7,600 to 15,000 pounds
Original price	$5,000

When equipped with the optional GM two-stroke diesel, the Oliver Super 99 could crank out an amazing 73.31 horsepower on the drawbar. Equipped with poppet-style exhaust valves, the engine had a unique wailing exhaust sound and had to be kept at high revolutions to produce power. *Ralph W. Sanders*

Years of production	1949–1956
Engine	Sheppard, 142 cubic inches (2.3 liters), 2 cylinders
Fuel	Diesel
Horsepower	25 belt (claimed)
Rated RPM	1,650
Transmission	4 speeds forward, 1 reverse
Weight	4,000 pounds
Original price	Unknown

Sheppard SD-2 Diesel

Built by the R. H. Sheppard Company in Hanover, Pennsylvania, Sheppard tractors were somewhat unique in the tractor industry in that from the beginning, they used a diesel engine of their own design to provide a high degree of power for very little money. The Sheppard was also the first direct-start diesel tractor built in the United States. The company claimed that its diesel engines burned half as much fuel as comparable gasoline-powered tractors. Adding to the frugality was the fact that diesel fuel cost about half as much as gasoline at the time.

The Sheppard tractors were equally simple, both in design and nomenclature. Customers were quick to understand that the SD-1 used a one-cylinder engine and was rated for a one-bottom plow, an SD-2 had two cylinders and a two-plow rating, and the SD-3 used a three-cylinder diesel to attain a three-plow rating. The company added an SD-4 in 1954, and, yes, it followed the same logic, having a four-cylinder engine. Regardless of the engine, the pistons, rings, bearings, sleeves, rocker arms, and injectors were all the same, so they could be interchanged easily. In fact, the commonality was so exact that one former owner recalls how in an emergency, one could cut the fourth cylinder off of the head gasket for an SD-4 model and use it for the SD-3 engine.

The irony is that one of the tractor components—a power steering unit that was also designed by Sheppard—became more valuable than the tractor itself. The power steering unit gained so much market share once it was applied to heavy-duty trucks and other equipment that Sheppard stopped building tractors in the mid-1950s to focus on manufacturing power steering units.

Reportedly, Sheppard produced approximately 1,943 tractors in the six years the SD Series was on the market. That included just 14 of the SD-1 models, 257 SD-2 models, 1,441 SD-3 models (which were obviously the most popular), and 231 SD-4 units.

Sheppard tractors were rather unique in the industry in that they were one of the first manufacturers to use a diesel engine of their own design in their tractors. The SD model number also revealed the engine size, which meant the SD-2 shown above had a two-cylinder engine. *Ralph W. Sanders*

Minneapolis-Moline GBD

It was in 1954 that Minneapolis-Moline announced it would offer diesel-powered tractors, as farmers were demanding bigger, more powerful tractors. Early offerings included the U and the UB diesels, along with the huge Model GTB-D. Standard features on the latter included a massive six-cylinder engine with a 4¼x5-inch bore and stroke and Lanova-design combustion chambers, which incorporated an energy cell directly opposite the injector.

Of course, the Minneapolis-Moline G Series actually started in 1938 when the Prairie Gold Model GT was introduced as standard-tread wheatland model. Features on that first G included full crown fenders, a 403-cubic-inch

Years of production	1955–1959
Engine	Minneapolis-Moline, 425.5 cubic inches (7.0 liters), 6 cylinders
Fuel	Diesel
Horsepower	62.78 belt (tested), 55.44 drawbar (tested)
Rated RPM	1,300
Transmission	5 speeds forward, 1 reverse
Weight	7,400 to 12,000 pounds
Original price	$4,320 (1955)

Minneapolis-Moline was known in later years for building high-horsepower wheatland-style tractors. One of the first was the massive GBD with its 425-cubic-inch six-cylinder diesel engine than provided nearly 63 belt horsepower. *Ralph W. Sanders*

four-cylinder engine, and a red grille that stood out against the yellow sheet metal. As the years went by, the model transitioned through the GTA and GTB models, gaining even more horsepower and more gears as the years went along. The GTB-D, of course, became the diesel model when it took on Minneapolis-Moline's first diesel engine and the company's first six-cylinder, while the GTC was an LPG-powered version of the GTB.

In 1955 the GTB Series was replaced with the GB tractors, including the GTD diesel version (shown). Each shared the same hood, grille, and engine with the comparable GTB model. The GT and GB Series were just the beginning of the monstrous G Series tractors to follow, eventually culminating in the massive 110-horsepower G1000.

Massey Ferguson 95 Super

As many will recall, Massey Ferguson initially got into the tractor market with small- and medium-horsepower tractors. Consequently, when farmers began to demand more horsepower and large wheatland tractors, Massey Ferguson was forced to turn to outside sources to fill the need.

As a result, the Massey Ferguson 95 Super is actually the same tractor as the Minneapolis-Moline G-VI, hence the Minneapolis-Moline engine under the familiar red and gray colors. Massey Ferguson had previously been marketing the Minneapolis-Moline GBD as the Model 95. However when the G-VI was introduced, the Massey Ferguson model became the 95 Super. Few American

Years of production	1959–1962
Engine	Minneapolis-Moline, 426 cubic inches (7.0 liters), 6 cylinders
Fuel	Diesel
Horsepower	78 PTO (claimed), 78 drawbar (claimed)
Rated RPM	1,500
Transmission	5 speeds forward, 1 reverse
Weight	8,335 to 12,695 pounds
Original price	Approximately $6,500

Faced with the need for a high-horsepower tractor for tillage applications, Massey Ferguson took the stop-gap route and rebadged the Minneapolis-Moline G-VI as its own Model 95 Super. *Andrew Morland*

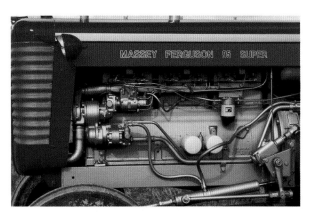

Equipped with a massive 426-cubic-inch six-cylinder Minneapolis-Moline diesel engine, the 95 Super provided nearly 75 PTO horsepower. Unfortunately for US farmers, the 95 Super was only offered in Canada and to export markets. *Andrew Morland*

farmers ever saw one though, as the 95 Super went only to Canada and export markets.

Even before the 95 and 95 Super, Massey Ferguson had offered the Massey Ferguson 98, which was actually an Oliver Super 99 (General Motors) in Massey Ferguson styling and paint. Similarly, the Massey Ferguson 97, which was the model that came after the 95 Super, was actually the Minneapolis-Moline 705/706 in red paint. It wasn't until the Massey Ferguson 1100 Series was introduced in 1964 that Massey Ferguson brought big tractor production in-house.

Years of production	1963–1969
Engine	Allis-Chalmers, 426 cubic inches (7.0 liters), 6 cylinders
Fuel	Diesel
Horsepower	D-21: 103.06 PTO (tested), 93.09 drawbar (tested)
	D-21 II: 127.75 PTO (tested), 116.41 drawbar (tested)
Rated RPM	2,200
Transmission	8 speeds forward, 2 reverse
Weight	9,500 pounds
Original price	$12,290

Allis-Chalmers D-21

The D-21 broke a lot of new ground for Allis-Chalmers when it was introduced in 1963. The tractor not only showed off modern styling elements, but more importantly, it was the first Allis-Chalmers tractor producing more than 100 horsepower. The 103-horsepower D-21 moved Allis-Chalmers into the big tractor market and helped erase the company's image as a small-tractor maker.

Although the D-21 was the last of seven D Series farm tractors, which began with the D-14 in 1957, it didn't look much like the rest of the D Series line. In contrast, the D-21 featured new, modern styling elements that were developed by a new industrial design department that had recently been created within the company. Perhaps its most eye-catching feature was the large rectangular grille with chrome trim that complemented the smooth, squared lines. It looked tough even when it wasn't hitched to anything.

At the same time, Allis-Chalmers marketed a new operator platform, advertised as being as big as a "ballroom." The platform was not only large by most tractor standards, but included full fenders that extended forward over the rear wheels to keep mud off the platform. On the D-21, the fenders were even designed to incorporate two field lights on each side for better visibility. Another unique feature was the suspended seat mounted on an inclined track. As the seat was moved forward, it was moved into a lower position, allowing the operator to find a more comfortable work position.

Unlike earlier models, the D-21 also featured a large 52-gallon fuel tank located behind the operator—enough to run nearly 10 hours at 75 percent load. For most farmers, that was more than enough to go a full day without refueling at lunchtime.

Last, but not least, the D-21 featured a new powertrain with an independent PTO. The constant mesh transmission offered four forward

The 103-horsepower D21 broke the 100-horsepower barrier for Allis-Chalmers and easily pushed the company into the big-tractor market. In addition to new styling not shared with the other D Series tractors, it came with a host of new features. *Tharran Gaines*

gears and one reverse gear. A two-speed gearbox doubled all gears to provide the operator with eight forward and two reverse gears.

Facing competition from other manufacturers as well as its own 190XT tractor, Allis-Chalmers introduced the D-21 Series II in 1965. Equipped with a new 3500 Series engine, it was basically the same tractor except it was now beefed up with a turbocharger that pushed the power rating to 127.75 horsepower at the PTO and 116.41 at the drawbar. This gave the D-21 II more than 15,000 pounds of pull, making it the largest tractor Allis-Chalmers had ever built.

Farmall 1206 Turbo

Years of production	1963–1967
Engine	International Harvester, 361 cubic inches (5.9 liters), turbocharged 6 cylinders
Fuel	Diesel
Horsepower	112.64 PTO (tested), 95 drawbar (tested)
Rated RPM	2,400
Transmission	8 speeds forward, 4 reverse
Weight	10,115 to 13,580 pounds
Original price	$9,450

When it was introduced in 1963, the Farmall 1206 became the first Farmall tractor to feature a turbocharger, hence a gold-scripted "Turbo" decal on each side of the hood. It was also the first row-crop tractor available with more than 100 horsepower. While the 361-cubic-inch diesel was the same one used in the smaller 806, the 1206 not only featured an exhaust gas turbine to drive a compressor, but also added a new lubrication system, a larger air cleaner and cooling fan, and boasted a stronger drivetrain. The turbocharger was built by Solar, a company based in San Diego that International Harvester had purchased a few years earlier.

International Harvester also added new flat-topped fenders to the 1206 that were later added as an option on the 706 and 806. While it was styled like the 706 and 806—muscle tractors in their own league—the 1206 incorporated a new paint scheme that included white grille, fenders, and wheels to set it apart. The 1206 also came with a welded, tubular grille that replaced the cast grille used on previous models.

Available options included the torque amplifier that was standard on earlier models, a factory cab, dual rear wheels, and the choice of a wide or narrow front axle.

Built from 1963 to 1967, the Farmall 1206 Turbo was the first International Harvester tractor to utilize a turbocharger. Ironically, the turbocharger itself was built by Solar, a San Diego–based company that International Harvester had acquired just a few years earlier. Thanks to its turbocharged engine, it also became the first row-crop tractor with more than 100 horsepower. *Randy Leffingwell*

Engine Modifications

For as long as tractors have been designed and built, engineers have constantly been looking at ways to get more power out of the tractor or engine. As we learned in the previous chapter, engineers often used bigger engines to put more power through the powertrain. However, there were also numerous times when tractor engineers simply took an existing engine and beefed it up in one way or another. Sometimes it was as simple as adjusting the governor to increase the rated engine speed. Other times it involved making an internal modification to the piston or the valvetrain.

In other cases, a tractor company simply borrowed the engine from an automobile or truck division, modified it to some extent, and put it into a tractor. That was the solution for both Case and International Harvester on a few occasions. However, one of the biggest changes came when tractor manufacturers started offering a diesel engine as an option as early as the 1940s.

By 1960, of the 25 tractors tested at the Nebraska Tractor Test Laboratory, 14 had diesel engines. Fifteen years later, in 1975, all 30 tractors that were tested in Nebraska were powered by diesels. Today, engine modifications continue to improve fuel efficiency and boost horsepower. Consider, for example, that the Massey-Harris Challenger with its 247.7-cubic-inch engine cranked out just 28.58 horsepower on the belt when it was introduced in 1936. Seventy-seven years later, Massey Ferguson released the Model 5613, which is considered to be a utility tractor, rated at 125 engine horsepower and 100 horsepower on the PTO. In comparison to the Challenger's 4.1-liter engine, the 5613 has a four-cylinder engine that is only slightly bigger at 4.4 liters.

The difference is today's model uses a turbocharged, intercooled diesel engine rated at 2,200 rpm that uses four valves per cylinder. In addition, it utilizes a high-pressure common rail direct-injection fuel system that feeds an injector at each cylinder. If that's not enough to think about, consider that today's automobiles are often powered by engines with a displacement of three liters or less, when six or eight cylinders were the norm just 30 to 40 years ago.

Ralph W. Sanders

Engine designers continue to modify engine components to crank more power out of fewer cubic inches, all while generally increasing the engine lifespan and decreasing the emissions. So let's take a look back at tractor engine evolution to see where we've been, and how we got to where we are today.

Rumely OilPull 30-60

Years of production	1910–1923
Engine	Rumely, 1,885 cubic inches (30.9 liters), 2 cylinders
Fuel	Gasoline
Horsepower	75.60 belt (tested), 49.91 drawbar (tested)
Rated RPM	375
Transmission	1 speed forward, 1 reverse
Weight	26,000 pounds
Original price	$2,700

The Advance-Rumely 30-60 OilPull was introduced in 1910 as the production version of the "Kerosene Annie," the first kerosene-powered tractor built by the M. Rumely Company as an experimental model. Like the original, it was an onerously heavy machine, weighing approximately 26,000 pounds—almost as much as the steam engines it was meant to replace. Even with 75 horsepower, it took nearly half of its power just to move the tractor, let alone any load. The first models were essentially variations of the company's steam engines converted to internal combustion; hence, they were still suited only for the largest farms and for threshing work.

In addition, the two-cylinder engine had such a large displacement (1,885 cubic inches) that it was too large to crank and, instead, had to be started with compressed air. Still, it was enough of a positive change to launch Rumely into an era of rapid growth.

The OilPull's biggest selling point, of course, was the fact that it ran on kerosene, which at the time sold for around seven cents per gallon. Compared to steam engines, it meant the operator no longer had to handle coal or wood or wait for steam to build up in the boiler. Better yet, there was no boiler to explode with devastating results.

On the other hand, it could be operated much more cheaply than a gasoline-powered tractor, since gasoline was more than twice the price of kerosene at 16 cents per gallon. According to company promotional material, that could amount to a difference of $4.50 per day—a large amount in 1910. Of course, the Advance-Rumely's ability to run well on kerosene,

As part of a demonstration conducted by Purdue University and the Indiana Agricultural College, three 30-60 OilPull tractors were coupled to a 50-bottom plow. Together, they turned over nearly 60 feet of soil at each pass, or an acre every 4½ minutes. Obviously, the demonstration caught the attention of everyone interested in power farming. *Ralph W. Sanders*

which had a much narrower temperature band, was attributed to two engine features. First, it was oil-cooled instead of water-cooled. Since oil has a much higher boiling point than water, this allowed the engine to run hotter, which permitted better combustion of kerosene. Key to temperature control was the large cooling tower at the front of all Rumely OilPull models. The exhaust was routed through the tower to create

Available as an option on some models, special extension rims proved to be valuable for breaking up marsh ground in certain parts of the country. *Ralph W. Sanders*

airflow, or air suction through the cooling unit without the use of a fan. The second thing that helped was water injection, which was controlled by the governor, to regulate combustion.

However, the 30-60 was rather slow in every sense of the word. There was only one forward speed at 1.9 miles per hour and one reverse speed. The engine's cycle was equally slow with a rated speed of only 375 rpm. That began to change, though, with Generation Two of the Advance-Rumely tractors. In addition to lighter weight, they featured a slightly higher engine speed and two forward speeds.

Although the Advance-Rumely OilPull 30-60 was in production for nearly 13 years, easy starting and improved performance eventually allowed gasoline, despite its higher price, to become the preferred fuel among farmers, ultimately bringing an end to all kerosene tractors.

John Deere Waterloo Boy N

Years of production	1917–1924
Engine	Waterloo Gasoline Engine, 465 cubic inches (7.6 liters), 2 cylinders
Fuel	Kerosene
Horsepower	25.97 belt (tested), 15.98 drawbar (tested)
Rated RPM	750
Transmission	2 speeds forward, 1 reverse
Weight	6,180 pounds
Original price	$1,050 (1921)

It's difficult for most people to think of a Waterloo Boy as anything other than John Deere's first successful farm tractor. But the fact is, the Waterloo Gasoline Engine Company, which was organized in 1895, as an engine manufacturer, actually developed and initially marketed the tractor. The Waterloo Boy had evolved from a self-propelled tractor designed in 1892 by John Froelich. Although the Froelich tractor never became a commercial success, it did form the base for the company's stationary, horizontal, two-cylinder engines—even though Froelich himself had since left the company.

In 1913, with the hiring of two new engineers, the Waterloo Gasoline Engine Company took a renewed look at powered tractors in response to changes in the marketplace and developed the Waterloo Boy One-Man Tractor. The company started with the L and LA, both of which featured a single-speed transmission. Both had an engine with a cylinder arrangement that was horizontally opposed. The L was a three-wheel model, while the LA was the same tractor with a wide front. Unfortunately, only a few models of each were ever sold.

Originally developed and marketed by the Waterloo Gasoline Engine Company, the Waterloo Boy became the tractor from which all John Deere two-cylinder tractors would evolve, following the purchase of the engine company by John Deere in 1918. *Ralph W. Sanders*

In the meantime, Waterloo Boy developed a new side-by-side, two-cylinder, four-cycle engine. This engine was then installed on an LA chassis to become the R. Like the L and LA, it still had only one forward gear, and steering was accomplished with a chain windlass steering system.

In 1917, the N was added. However, the latter was equipped with a two-speed gearbox instead of the single speed and featured a worm and sector–type steering. The largest of the models, it offered 12 horsepower on the drawbar and 25 on the belt with the engine running at only 750 rpm on kerosene.

As most everyone knows however, the Waterloo Boy tractors soon attracted the attention of John Deere, which had serious ambitions to move into the rapidly growing tractor market. Not only had John Deere been unsuccessful at developing its own tractor model, but company executives reasoned that purchasing an established product would be a shortcut in developing both a unit to sell and an established market. Buying the Waterloo Gasoline Engine Company would not only solve those issues, but provide Deere & Company with its own engine factory.

The Deere & Company acquisition was completed in 1918 at a cost of $2.35 million for the entire Waterloo Gasoline Engine Company. The N remained in production with only minor changes until 1923—carrying the Waterloo Boy name the entire time. It wasn't until late 1923 when Deere introduced the D, calling it a 1924 model at the time, that the reign of the Waterloo Boy came to an end.

Years of production	1917–1922
Engine	International Harvester, 283.7 cubic inches (4.6 liters), 4 cylinders
Fuel	Kerosene
Horsepower	16.52 belt (tested), 11.00 drawbar (tested)
Rated RPM	1,000
Transmission	3 speeds
Weight	3,660 pounds
Original price	Unknown

International Harvester 8-16 Junior

Like a number of tractor manufacturers in the 1900s, International Harvester was formed as a merger of two major farm equipment companies that were forced to become partners in order for both to survive. It happened to Holt and Best when the former enemies reluctantly joined forces to form Caterpillar in 1925.

The merger of the McCormick Harvester Company and the Deering Company several years earlier, in 1902, was no more welcomed by company executives and dealers. In effect, the two merged companies continued to operate much as they had before, to the point of developing two different tractor lines—Titan and Mogul. Most of those early tractors, however, were designed for plowing and belt work, leaving jobs like cultivation, planting, and mowing hay to a team of horses.

In an effort to mechanize those tasks, too, companies began looking at small tractors by the 1910s. Mogul came out with their answers in the form of an 8-16 and 12-25, while Titan introduced the 10-20 and 12-25. It wasn't until 1917 that the Mogul and Titan tractor engineers were pulled together into one department with a single goal. One factor that prompted the merger was the new Fordson tractor. Introduced the same year as International Harvester's engineering consolidation, it was small and affordable and was

Often referred to as the "kerosene tractor," the International 8-16 Junior proved there was a market for a powerful, lightweight tractor, ultimately paving the way for the Farmall models that followed. Its overhead-valve engine with a bore and stroke of 4x5 inches provided 16 horsepower on the belt. *Ralph W. Sanders*

proving to be serious competition for all tractor manufacturers, including International Harvester.

One of the first tractors to come out of that joint effort was the International 8-16 Junior, not to be confused with the 8-16 Mogul introduced in 1914. Often referred to as the "kerosene" tractor, the 8-16 Junior was a small tractor that was designed after the existing International

The 8-16 not only used the same engine as the G International truck, but it also borrowed the same style of sloping hood and the same fan-cooled radiator. *Ralph W. Sanders*

trucks. Not only did it have the same style of sloping hood, but it used the same engine as the Model G International truck. The 8-16 also had a three-speed transmission—an unusual feature at a time when one and, at most, two speeds were the norm—and a fan-cooled radiator, which actually was positioned behind the engine.

Case 10-18 Crossmotor

Years of production	1918–1920
Engine	J. I. Case, 235.9 cubic inches (3.9 liters), 4 cylinders
Fuel	Kerosene
Horsepower	18.14 belt (tested), 11.24 drawbar (tested)
Rated RPM	1,050
Transmission	2 speeds forward, 1 reverse
Weight	3,760 pounds
Original price	Unknown

You might say J. I. Case had a method to its madness when it mounted the engine crossways in the Model 10-18 Crossmotor and other Crossmotor tractors that followed. To this point, most tractors used a riveted frame that often flexed and bent in the field, often pulling transmission gears out of alignment.

Case figured if they mounted the engine crossways, they could build a shorter, more rigid frame with the engine in the middle to balance the weight. Unfortunately, Case hadn't had much luck with internal combustion engines, having tried various two-cylinder designs. So it was almost by luck that the engineers turned to the automobile side of the business for their Crossmotor engine.

Case had just recently acquired the Pierce Motor Company, and it was discovered that one of the company's four-cylinder, overhead-valve engines could be reworked for use in a tractor. Hence, the Case Crossmotor line was born. Other benefits of the Crossmotor design included better access to the engine for simple repairs and the ability to install a fully enclosed, two-speed transmission. Finally, belt-driven implements like threshers, grinders, mills, and so on could be driven directly off the end of the crankshaft.

The Crossmotor line actually started with a three-wheel model, similar to other tractor designs of the time. Branded as the 10-20, it was driven by a single rear wheel most of the time except when the other wheel clutched into the powertrain when extra pull was needed.

Fortunately, the company began introducing more conventional four-wheel models that proved to be more popular. Among these was the 9-18,

Despite being a recognized name in the steam engine industry, Case had a little more trouble establishing itself in the kerosene tractor arena. That is, until a suitable engine practically fell in the design engineers' laps. Mounting it crosswise, though, made the real difference for nearly a decade. *Ralph W. Sanders*

which was later upgraded in 1919 to become the 10-18 (shown). Like the 9-18, it featured a cast-iron frame that included the engine block. The engine also included a few rather advanced features, including a water pump and a sight gauge so that circulating oil could be easily checked.

Eventually, the 10-18 itself was updated as the 12-20 with a new disc clutch that replaced the internally expanding type. All together, Case utilized the Crossmotor concept for a total of 13 years.

Years of production	1934–1937
Engine	Waukesha, 196.1 cubic inches (3.2 liters), 4 cylinders
Fuel	Gasoline, kerosene, or distillate
Horsepower	24.22 belt (tested), 17.98 drawbar (tested)
Rated RPM	1,275
Transmission	5 speeds forward, 1 reverse
Weight	3,450 pounds
Original price	Unknown

Minneapolis-Moline JT

The JT was one of the first major redesigns to carry the Minneapolis-Moline name after the three-way merger of Minneapolis Steel and Machinery, the Minneapolis Threshing Machine Company, and the Moline Plow Company in 1929. Due to the merger, the line temporarily consisted of both Minneapolis and Twin City tractor models. However, the Minneapolis line was discontinued as inventories were sold out, leaving the Twin City brand to carry on. Those included the KT, MT, and FT models that were built until 1934.

It was at that time Minneapolis-Moline introduced the two-plow Universal JT. Starting with the earlier MT model, Minneapolis-Moline had

Equipped with five forward speeds (four field speeds and a high-road speed) and a variable speed governor controlled from the seat, the Minneapolis-Moline Universal J could travel at any speed between a crawl and 12.2 miles per hour. The JT was also ahead of its time with adjustable tread wheels. *Ralph W. Sanders*

started using the name "Universal" to identify its row-crop tractors—or as the company's early advertisements boasted, tractors "suited to all purposes and conditions and adapted to a great variety of users."

Additional models included the standard-tread JTS and a JTO orchard model. All three featured a new 196-cubic-inch engine with an "F" head configuration. This placed the exhaust valves in the cylinder head and the intake valves in the block with the idea that it provided more water jacket space around the exhaust valves for improved cooling. It was advertised that customers would also appreciate "the patented 3-fuel manifold, which efficiently handles gasoline, kerosene and distillates without water injection."

Another important feature was the five-speed transmission that provided more working speeds and a high road speed of 12.2 miles per hour. A third unique feature was a straight-through rear axle with sliding rear hubs that allowed tread adjustment to match a wider variety of row crops, particularly with the larger rear wheels. The fenders even adjusted in and out with the wheels to protect the operator from dust and debris. As a result, the tractor was claimed to plow without side draft and cultivate two or four rows at a time.

The new "F" head configuration on the JT placed the exhaust valves in the cylinder head and the intake valves in the block with the idea that it provided more water-jacket space around the exhaust valves for improved cooling. *Ralph W. Sanders*

Massey-Harris Challenger "Twin-Power"

Having absorbed in 1928 the old Wallis tractor brand that included the Wallis 10/20, Massey-Harris decided it was time for an update in 1936 to a more modernized row-crop tractor. So along with the new conventional standard Pacemaker, the company introduced the Challenger—

As the first row-crop tractor for Massey-Harris, the Challenger was unique in that it offered "Twin-Power." In effect, it provided two engine speeds or governor settings. One was for drawbar work, while the other was for belt applications. *Ralph W. Sanders*

Years of production	1936–1938
Engine	Massey-Harris, 247.7 cubic inches (4.1 liters), 4 cylinders
Fuel	Gasoline or distillate
Horsepower	28.58 belt (tested), 20.03 drawbar (tested) 40.73 belt (tested), 29.83 drawbar (tested)*
Rated RPM	1,200, 1,400**
Transmission	4 speeds forward, 1 reverse
Weight	3,900 pounds
Original price	$1,055 (1938)

* Twin-Power model after 1938
** Belt setting

Massey-Harris's first row-crop tractor. Both models continued to use the old boilerplate frame developed by Wallis and used a 248-cubic-inch engine with about 29 horsepower. Steel wheels were standard, as was a narrow, tricycle-style front axle on the Challenger. However, rubber tires and an adjustable wide front axle were both optional. The tractor could also be equipped with an engine-driven mechanical implement lift.

However, things changed in 1938 when both tractors received styled grilles and sheet metal and bright red paint with yellow wheels. At the same time the tractors got the designation of Twin-Power models. Basically, Twin-Power provided two engine speeds, or governor settings, to match different types of jobs. For drawbar work, the operator used the 1,200-rpm setting, supposedly to reduce the amount of wheel spin. A higher 1,400-rpm setting was reserved for belt work. In fact, an interlock on the transmission prevented the operator from using the higher engine speed in the field.

When combined with the use of gasoline, instead of distillate, the Twin-Power Challenger turned out more than 40 horsepower on the belt, even though it was initially rated at 20 on the drawbar and 32 on the belt. As a result, it turned out to be a very popular model in both the wheat belt and among row-crop producers.

Allis-Chalmers WD-45

The WD-45 was much more than the first Allis-Chalmers tractor available with power steering. At the time it was introduced, in 1953, it was also the most powerful tractor of its size. Much of the credit goes to a new Power Crater gasoline engine that, according to the Nebraska Tractor Test Laboratory, gave the WD-45 about 25 percent more power than the previous WD. In addition to an extra ½ inch of stroke, the concave piston crown, or Power Crater, allegedly increased turbulence in the cylinder and raised the compression ratio to 6.45:1.

When the redesigned engine was combined with the proven Traction Booster system, the field results were impressive. The WD-45 quickly became one of the most popular tractors in the Allis-Chalmers line, as no tractor in its weight and power class could match it in performance.

The WD-45 introduced more than just power and performance, though. It was also the first tractor on the market to offer a hitch that could be coupled from the tractor seat. Called the Snap-Coupler, it required the operator to simply back the tractor over the implement tongue and listen for the "snap," which meant the hitch had engaged.

In 1954, a diesel model was added, and in 1956, the WD-45 became the first Allis-Chalmers farm tractor to have power steering. Even when it was replaced in 1957 by the D-17, the WD-45 remained one of the most popular tractors Allis-Chalmers ever built.

Years of production	1953–1957
Engine	Allis-Chalmers, 226 cubic inches (3.7 liters), 4 cylinders (gasoline or LPG)
	230 cubic inches (3.8 liters), 6 cylinders (diesel) after October 1954
Fuel	Gasoline, LPG, or diesel (after 1954)
Horsepower	29 belt (claimed), 23 drawbar (claimed)
Rated RPM	1,400
Transmission	4 speeds forward, 1 reverse
Weight	3,995 to 4,285 pounds
Original price	$2,380 (1956)

Thanks to its new Power Crater gasoline engine, the Allis-Chalmers WD-45 was the most powerful tractor of its size when it was introduced in 1953. It also became the first tractor on the market to offer a hitch that could be coupled from the tractor seat. *Ralph W. Sanders*

Allis-Chalmers D-19

When Allis-Chalmers introduced the D-19 in 1961, the idea was to acquire a portion of the market for tractors over 60 horsepower that the previously released D-17 could not reach. In the process, however, the D-19 did a lot more than put Allis-Chalmers in the five-plow tractor market. It also established a number of firsts in the tractor industry, including several in the engine.

Equipped with a six-cylinder, 262-cubic-inch engine powered by gasoline, diesel, or propane, the diesel version became the first tractor in the industry with a standard equipment turbocharger. The D-19 was also the first tractor tested at the Nebraska Tractor Test Laboratory with a dry air cleaner.

The D-19 was originally slated to be the D-18. The problem was that at 60 horsepower, the D18 was less than one plow bottom larger than the D-17, or approximately 15 percent more powerful. That wasn't enough to entice owners to trade up and it wasn't enough to compete in the power race, which had just been bumped up by the company's major competitors. At nearly the last minute, the company decided to increase the D-18's horsepower and give it a new number.

There wasn't a problem moving the G262 six-cylinder gasoline engine up to the needed horsepower, which was now based on a range of 65 to 70 horsepower, but the naturally aspirated diesel wasn't capable of reaching that horsepower level at the required 2,000 rated rpm. The engineers were basically left with two options on the diesel model—bore the 262-cubic-

Years of production	1961–1964
Engine	Allis-Chalmers, 262 cubic inches (4.3 liters), 6 cylinders
Fuel	Gasoline or distillate
Horsepower	71.54 PTO (tested), 63.91 drawbar (tested)
Rated RPM	2,000
Transmission	Power Director with 8 speeds forward, 2 reverse
Weight	6,650 pounds (gas), 6,840 pounds (diesel)
Original price	$5,300 (gasoline), $6,080 (diesel)

The D-19 was not only one of Allis-Chalmers' most popular models during the 1960s, but it was also a pioneer in many respects. The diesel model was the first in the industry to use a turbocharger. It was also the first to offer power steering as standard equipment. *Andrew Morland*

inch engine out to 290 inches or add a turbocharger. The turbocharger was ultimately selected, changing tractor design forever.

One more first on the D-19 was the fact that power steering was standard. It also used a new self-energizing brake system that required very little force at the pedal to provide impressive braking during turns and stops.

Within the Allis-Chalmers line, the D-19 also represented a departure from previous designs with its straight-through rear axle equipped with sliding hubs and a choice of 34- or 38-inch rear tires. Up to this point, Allis-Chalmers had the smallest tires in the industry. In the meantime, the Power-Director three-position hand clutch and four-speed constant-mesh transmission were similar to the D-17, providing high-low, on-the-go shifting, continuous PTO, and eight forward and two reverse speeds.

Oliver 1950

When the Oliver 1950 appeared in 1964, it had exactly the same engine that the previous 1900 model used, except the 1950 had more than a 5 percent increase in power. At its new 105-horsepower rating, it became the most powerful row-crop tractor in the industry at the time.

Of course, Oliver had already seen a similar increase in horsepower when it introduced the 1900 Series B to replace the original 1900. In all cases, the company used the General Motors 4-53 four-cylinder, two-stroke, 212-cubic-inch diesel engine. On the 1900, it was rated at 2,000 rpm and 89.35 horsepower. At 2,200 rpm on the Series B, it went to 98.54 horsepower.

However, if a little is good, more must be better, so Oliver wound the General Motors engine on the 1950 all the way to 2,400 rpm, establishing a new level for horsepower. What's more, Oliver didn't just advertise the horsepower but guaranteed it. With the introduction of the 50 Series, Oliver introduced its "Certified Horsepower" program that called for a decal on each tractor that guaranteed it to deliver a specified minimum horsepower. The same decal, with a change in the brand name, appeared on the comparable Cockshutt models as well.

Years of production	1964–1967
Engine	General Motors, 212.4 cubic inches (3.5 liters), 4 cylinders (4-53 two-cycle with blower)
Fuel	Diesel
Horsepower	105.8 PTO (tested), 99.27 drawbar (tested)
Rated RPM	2,400
Transmission	6 speeds forward, 2 reverse (gear)
	18 speeds forward, 6 reverse (Hydraul-Shift)
Weight	10,220 pounds
Original price	$12,700

With 105 PTO horsepower, the Oliver 1950 not only became the most powerful row-crop tractor at the time, but it was unique in that it came with a guarantee. Like the other 50 Series Oliver tractors, along with the comparable Cockshutt models, it carried a "Certified Horsepower" decal that guaranteed it to deliver the specified minimum horsepower. *Heritage Iron*

Of course, it's easy for Cockshutt to make that promise, too, since the Cockshutt 1950 and Oliver 1950 were exactly the same tractor, except for the paint and nameplate. After the White Motor Company acquired Cockshutt in 1962, the tractors built in Brantford, Ontario, were discontinued and replaced with Oliver tractors painted Cockshutt colors.

Both the Oliver model and Cockshutt version of the 1950 were available with an adjustable front axle or front-wheel assist, which first became available as an option on the 1900 B Series. Other options included a factory-installed cab, the Hydraul-Shift transmission and fender tanks, which used fuel tanks incorporated into the fenders to increase fuel capacity from a meager 40 gallons to 118 gallons.

Something Special

Special-edition tractors may not have been as common as limited-edition or special-edition automobiles, but that doesn't mean they didn't exist. Some were designated by little more than an emblem to signify an anniversary, while a couple of others were adorned in red, white, and blue paint to celebrate the bicentennial of the United States.

Other tractor brands or models were never intended to be limited editions, but turned out to be in production for as little as two years, making them special in their own way. For tractor collectors, it simply makes those models that much more elusive and valuable.

There are probably other special-edition models or extremely limited productions that we've missed. And that doesn't even count the dozens, if not hundreds, of one-of-a-kind prototypes that tractor companies built over the years. Many of those never even made it out of the test lab or, at the most, were built in quantities of three or four as test subjects. Some of them were revised and introduced as a new model with the appropriate changes, while others never saw the light of day.

A good example of the latter is the Allis-Chalmers 8095. It was one of those tractors remembered only in history, as one that never went into production. Scheduled for release in 1986, the 8095 would have been the largest rigid-frame tractor ever produced by Allis-Chalmers. While it looked a lot like the existing Allis-Chalmers 8000 Series tractors, it had a larger cab and a longer wheelbase, and was powered by a 643-cubic-inch, six-cylinder Komatsu diesel engine (now obsolete) that was derated to provide 225 PTO horsepower.

Of course, the nail in the coffin for the Allis-Chalmers 8095 was the purchase of Allis-Chalmers Corporation by Klöckner-Humboldt-Deutz in 1985, which brought the production of all current and future Allis-Chalmers tractors to a grinding halt.

A similar story could be told about the John Deere C, built in 1927 and 1928—even if it wasn't the victim of a company buyout. The C was actually a predecessor or prototype to the GP and was designed to compete with the Farmall row-crop tractor that had been introduced a few years earlier. In effect, the C was called a row-crop tractor because it was made to straddle the rows with a higher stance than the D. Prior to the fall 1928 production, after fewer

Ralph W. Sanders

than 200 models had been built, the decision was made to rename the tractor as the GP. Most 1927 models were recalled and rebuilt, incorporating the minor changes that had been added to the GP.

Whether a model was built in limited numbers by design or by fate, here are some of the most noteworthy special editions and limited productions that actually made it to the dealership and farmers' hands.

Advance-Rumely 6A

Years of production	1930–1931
Engine	Waukesha, 404.3 cubic inches (6.6 liters), 6 cylinders
Fuel	Diesel
Horsepower	48.37 belt (tested), 33.57 drawbar (tested) at 1,350 rpm
Rated RPM	1,200 and 1,365
Transmission	6 speeds forward, 2 reverse
Weight	6,370 pounds
Original price	Unknown

The Advance-Rumely Model 6A certainly wasn't a special tractor, even though it was totally different than anything Advance-Rumely had manufactured to that point. However, it was rather limited in number, as only 802 units were built in 1930 and 1931.

Through the first part of the 20th century, Advance-Rumely was known for its OilPull tractors, which were particularly suited for plowing and belt work. But times were changing, especially with the introduction of the Fordson tractor. Farmers were looking at smaller tractors that could do more jobs on the farm, and gasoline-powered tractors were becoming more popular due to easier starting and the tendency to offer more power.

The company's first venture into an all-purpose tractor was the Advance-Rumely Do-All, introduced in 1928. However, the company soon followed with the 6A in 1930 as a more conventional standard-tread model. In fact, it was advertised as "a four-plow tractor at the weight of a three" and "a six-cylinder at the price of a four—and six speeds forward."

Powered by a 404-cubic-inch, six-cylinder Waukesha engine coupled to a three-speed transmission, it would seem to offer just three speeds forward. However, by rating the engine at both 1,200 and 1,360 rpm, the company was able to advertise six speeds, even if it was a technicality.

The 6A was rather short-lived, though. As the Depression worsened in the early 1930s, Advance-Rumely found itself running out of money and looking for a source of funding. After several months of negotiation, Advance-Rumely was acquired by Allis-Chalmers in May 1931, at which time the

As one of the only conventional tractors built by Advance-Rumely, the 6A featured six forward speeds, compared to the usual three offered by most competitors. It also proved capable of pulling 67 percent of its own weight. Unfortunately, production only lasted one year before Rumely was acquired by Allis-Chalmers. *Ralph W. Sanders*

6A was met with internal competition from the U and E 25-40 being built by Allis-Chalmers. It's also been reported that there were more than 700 Advance-Rumely 6A tractors left in inventory when Allis-Chalmers took over the company. Hence, the 6A, along with the Do-All, were dropped from the lineup. Allis-Chalmers did continue to offer unsold inventory of the 6A until at least 1934, when it was still listed in the Allis-Chalmers catalog.

Graham-Bradley 104

The Graham-Bradley tractors weren't really special-edition tractors, but they still hold a special place in history, as they were only built in two models for two years and marketed by Sears Roebuck. The tractors

Years of production	1938–1940
Engine	Graham-Paige, 218 cubic inches (3.6 liters), 6 cylinders
Fuel	Gasoline
Horsepower	25.2 drawbar (tested)
Rated RPM	1,500
Transmission	4 speeds forward, 1 reverse
Weight	Unknown
Original price	Unknown

were advertised as "built by Graham, equipped by Bradley, guaranteed by Sears." The tractors not only had the strength of Sears Roebuck behind them, but they were also built on the reputation of the Graham-Paige Motors Corporation in Detroit, which was known for its Graham-Paige automobiles. Unfortunately, automobile sales were beginning to slump in the late 1930s, which prompted the Graham brothers to look into tractor manufacturing.

Only one base model was offered, but it was sold as two model numbers. The 103 was the row-crop configuration with a tricycle front end, while the 104 was the standard-tread model. Both were marketed through the Sears Roebuck catalog as well as Graham-Paige automobile dealerships. Rated as a two-plow tractor, the Graham-Bradley used the same engine as the Graham-

Even without the nameplate, it's easy to recognize Graham-Bradley's unique design, which is set off by its sleek sheet metal and chrome accents. Marketed by Sears Roebuck, the Graham-Bradley was only produced for two years in two models. *Ralph W. Sanders*

Paige automobiles. In this case it was a Graham-Paige six-cylinder 214-cubic-inch engine limited to 1,500 rpm.

As one of the most streamlined tractors on the market, the Graham-Bradley featured an enclosed engine and silver accents on both the contoured grille and the engine side panels. The belt pulley was rather unusual for the time as well, since it was driven from the rear of the transmission. By locking out the final drive via a lever, the operator could select any transmission speed, including reverse, for the belt pulley. Rubber tires were also standard, and the adjustable rear wheels could be moved from 56 inches to as wide as 84 inches.

Unfortunately, sales were disappointing, and the last Graham-Bradley tractor was sold in 1940 with production limited to only 300 tractors. A year later in 1941, the factory was converted over to war products, including aircraft and marine engine parts. Three years later the factory was sold to Kaiser-Frazer Corporation, while the Graham brothers went into real estate.

Farmall C (1950 Demonstrator)

In 1950, International Harvester launched a sales program for its smallest tractors, in which a limited number of Farmall Cubs, Super A models, and C models were painted white as sales demonstrator units for the company's "Mid Century" sales promotions. The theory was that nothing would catch the farmer's attention more than a bright white tractor. In addition, wheel inserts boasted sales slogans: "Try it now. You'll like it all the time." and "More than a Million Farmall Users Know Why You'll Like a Farmall 'C'." Although many of these tractors were later repainted red after the promotion was over, a few escaped, which has made them very valuable among collectors.

Conversely, a few people have since painted tractors white in an attempt to add value to the tractor or to simply add a replica to their own collection. However, for those who started to sand a tractor in preparation for restoration and discovered white paint underneath, it's been the equivalent of winning the lottery.

Years of production	1948–1951
Engine	International Harvester, 113.1 cubic inches (1.9 liters), 4 cylinders
Fuel	Gasoline
Horsepower	21.12 belt (tested); 18.57 drawbar (tested)
Rated RPM	1,650
Transmission	4 speeds forward, 1 reverse
Weight	2,845 pounds
Original price	$1,500 (standard model)

While the Farmall C was produced by the thousands and found its way onto a wide variety of operations, a special group of "demonstrator" models that were painted white for a sales promotion are still considered rare and highly collectible. *Randy Leffingwell*

It didn't take a lot of gimmicks to sell the C, though. During its time, it earned a prominent position at county and state fairs. The Farmall C basically replaced the B, and even though it used the same 113-cubic-inch engine, it was rated at 1,650 rpm, instead of 1,400 like the B. That gave it three more horsepower. It also featured a higher operator platform with a seat and steering wheel that were centered on the platform, giving the operator a better view, especially when using a cultivator—one of the C's specialties. During its four years of production, International Harvester sold nearly 80,000 copies of the C before replacing it with the Super C.

Ford Jubilee (1953)

The Ford NAA, which was introduced in 1953, represented a major change in the Ford line from the previous N Series models, while introducing a number of new features. Although all NAA models are often referred to as a Jubilee, the Jubilee designation was technically used in 1953 *only* to celebrate Ford's 50th anniversary as a company. In fact, the first year's models carried a circular emblem on the grille that read "Golden Jubilee Model 1903–1953."

The word "jubilee" has biblical significance, as mentioned in the book of Leviticus, which states that a jubilee year is to occur every 50th year,

Years of production	1953–1954
Engine	Ford, 134 cubic inches (2.2 liters), 4 cylinders
Fuel	Gasoline
Horsepower	31.14 belt (tested), 26.8 drawbar (tested)
Rated RPM	2,000
Transmission	4 speeds forward, 1 reverse
Weight	2,550 to 4,392 pounds
Original price	$1,560 (1954)

Although it was technically part of the new Ford NAA line, the Ford Jubilee was actually produced in 1953 for one year only to commemorate Ford's 50th anniversary. The remaining year of NAA production in 1954 was distinguished by a different hood emblem. *Ralph W. Sanders*

during which slaves and prisoners would be freed, debts would be forgiven, and the mercies of God would be particularly manifest. However, a golden jubilee is also described as a royal ceremony to celebrate a 50th anniversary, particularly in other countries, including the United Kingdom, Korea, and China, where jubilee celebrations date back to 91 BCE.

The nose emblem on the next year's 1954 models had a similar appearance to the Jubilee models, except the words referring to the anniversary were replaced with stars that circled the outside border.

Among the changes made on all NAA models was a new vane-type hydraulic pump that was powered off the engine. This also meant the advent of live hydraulics, since hydraulic power no longer depended on engagement of the PTO. The new styling also gave the Ford NAA a look that differed from the previous N Series tractors as well as the Ferguson TO Series tractors that Ferguson was continuing to build.

The NAA allowed Ford to market a larger, more advanced tractor than the 8N. For starters, it was 4 inches taller, 4 inches longer, and around 100 pounds heavier than the 8N. It was equipped with a larger, more powerful overhead-valve "Red Tiger" engine. The new inline four-cylinder engine featured a 134-cubic-inch displacement rated at 31 horsepower.

Among the NAA's other improvements, beyond the extra power and hydraulics, were a better governor and a temperature gauge on the instrument panel. The NAA, or Jubilee, also had the muffler relocated under the hood alongside the engine, where it reduced the chance of a fire in hay fields and stubble.

Cockshutt Black Hawk 50

The Cockshutt Black Hawk 50 was just one of several Black Hawk tractors built in 1956 to commemorate the company's purchase of the Black Hawk equipment line from the Ohio Cultivator Company. While each of the tractor models that carried the Black Hawk label was built over a span of years, only the 1956 models received the special designation. In effect, Cockshutt replaced the Canadian "Cockshutt Deluxe" label with a bold "Black Hawk" decal. The tractors also had more standard features, including a cigarette lighter and a two-tone paint scheme consisting of a classy red chassis with yellow sheet metal and wheels.

As for the base model itself, the Cockshutt 50 had the same powertrain as the 40, except the 50 used a Buda 273-cubic-inch engine, as opposed to the 230-cubic-inch Buda engine in the 40. An independent PTO and hydraulic system with remote cylinder were optional. The 50 also had the choice of adjustable or nonadjustable front axles.

The 40 and 50 were among the last Cockshutt tractors to use the Buda engine. After Allis-Chalmers acquired the Buda Company in 1953, Cockshutt saw no reason to help a competitor and turned to Perkins and Hercules engines for power with the subsequent series.

Years of production	1953–1957 (Black Hawk in 1956 only)
Engine	Buda, 273 cubic inches (4.5 liters), 6 cylinders
Fuel	Gasoline or diesel
Horsepower	55.56 PTO (tested), 51.59 drawbar (tested) (gas) 51.05 PTO (tested); 46.22 drawbar (tested) (diesel)
Rated RPM	1,650
Transmission	6 speeds forward, 2 reverse
Weight	5,532 pounds
Original price	Unknown

The Black Hawk 50 was just one of several Cockshutt tractors built in 1956 to commemorate the company's purchase of the Black Hawk equipment line from the Ohio Cultivator Company. *Gary Alan Nelson*

Cockshutt Golden Arrow

Years of production	1957
Engine	Hercules, 198 cubic inches (3.2 liters), 4 cylinders
Fuel	Gasoline
Horsepower	38 PTO (claimed), 33 drawbar (claimed)
Rated RPM	1,650
Transmission	6 speeds forward, 2 reverse
Weight	4,154 pounds
Original price	Demonstrator models

The Cockshutt Golden Arrow was a promotional model that the company introduced in 1957 to demonstrate its new draft-sensing three-point hitch system. To show off the system, Cockshutt basically put the new hitch, rear end, and a new transmission onto a Model 35 and equipped it with the same Hercules four-cylinder 198-cubic-inch gasoline engine. Only 135 Golden Arrow models were built, and the model was never tested at the Nebraska Tractor Test Laboratory. However, it was rated at 33 horsepower on the drawbar.

One year later, in 1958, the Golden Arrow was restyled and revised to become the Cockshutt 550, which had identical specifications. It's been said

The Golden Arrow model was another of the special-edition tractors built by Cockshutt in the 1950s. In this case, it was a promotional model to demonstrate the new draft-sensing three-point hitch system. Only 135 models were built, with many of them returned to the factory. *Ralph W. Sanders*

that Cockshutt originally intended to recall all the Golden Arrow models and rebuild them into Model 550 tractors, as well. However, many were never returned to the factory, making those units rather collectible today.

Steiger Panther II "Spirit of '76"

The 1970s was a prime time for the sale of four-wheel-drive tractors in North America, as both farms and implements were growing in size. Steiger had already been in business since the 1960s and was a well-respected name in the industry.

So, Steiger saw the opportunity to celebrate America's bicentennial in

Years of production	1974–1976
Engine	Cummins, 855 cubic inches (14.0 liters), 6 cylinders
Fuel	Diesel
Horsepower	310 engine (claimed), 262 drawbar (claimed)
Rated RPM	2,100
Transmission	4 speeds
Weight	Unknown
Original price	$43,200

To commemorate America's bicentennial and its last year of production, 50 Steiger Panther II models—one for each state in the US—received a special paint job and decals and were issued as "Spirit of '76" models. *Toy Tractor Times*

1976 with a special tractor sporting a red, white, and blue paint scheme. For Steiger, it was a Panther II model that was also decaled as the "Spirit of '76," although it was often called the "Stars and Stripes" Steiger as well. All other tractors at the time were painted Steiger green with a black grille and trim.

The 1976 model year was also the last for the Panther II before it became the Panther III. In addition, Steiger built only a very limited number of the tractors—50 total, or one to recognize each state in the United States. The tractors were actually built in September 1975. Delivery then took an additional six weeks, because the finished tractors were sent unpainted to a body shop in Fargo, North Dakota, where Steiger was based, to have the red, white, and blue paint scheme added. Even after they were shipped, the dealer received an instruction flyer showing them where to place the star decals.

One of the first Spirit of '76 models was sold by Hoober, Inc., to a farmer named Norris Hayman, who farmed in eastern Maryland. More than 25 years later, Hayman was still using the tractor on his farm.

The Panther II was a lot more than a patriotic model, though. As one of the most powerful tractors in the Steiger lineup at the time, it featured an 855-cubic-inch or 14-liter Cummins diesel engine that boasted 310 horsepower and 10 forward speeds from the transmission.

Ten years later, in 1986, following hard times by virtually all farm equipment manufacturers, the Steiger Tractor Company was purchased by Case IH. Shortly thereafter, the tractor color changed from lime green to Case IH red.

Case 1570 "Spirit of '76"

Years of production	1976
Engine	J. I. Case, 504 cubic inches (8.3 liters), 6 cylinders
Fuel	Diesel
Horsepower	180 PTO (claimed), 152 drawbar (claimed)
Rated RPM	2,100
Transmission	12 speeds forward, 4 reverse
Weight	13,000 pounds
Original price	$30,000 (1978)

It would be enough to say the Case 1570 was the largest, most powerful two-wheel-drive tractor of its time when it was introduced in 1976. With its turbocharged version of the Case 504-cubic-inch, six-cylinder direct-injection diesel under the hood, it produced 180 horsepower at 2,100 rpm. That made it even more powerful than the four-wheel-drive Case 1470 Traction King.

However, the 1570 had even more going for it. Since it was being introduced during the 200th anniversary of the United States, Case decided to take advantage of the country's patriotism and paint the 1976 production run with a special red, white, and blue paint scheme that included the nation's stars and stripes. Of course, it was appropriately named the "Spirit of '76."

The 1570 also used the Case "Agri-King" name, which was first used with the 70 Series. In addition to having a cab as standard equipment, the 1570 also featured a 12-speed transmission. All total, the 1570 was built from 1976 until 1978.

This wasn't Case's first attempt at a special-edition tractor. The company was also known for its "Black Knight" edition, which was a Case 870 with a unique black, silver, and red paint scheme. Needless to say, both the Black Knight and Spirit of '76 have appeared in numerous rural America parades—both as new tractors and enviable restorations.

Case issued its own "Spirit of '76" model in 1976. However, this time, it was part of the first year of production of the new Model 1570—the most powerful two-wheel-drive tractor of its time. *Ralph W. Sanders*

Put It in Four-Wheel Drive

Ever since the day tractors replaced horses on North American farms, manufacturers have been looking for ways to put more power to work. This, of course, led to larger tractors with more powerful engines that could pull bigger equipment. However, by the late 1950s and early 1960s, farmers were in need of more than just power. This actually prompted some to go to the extreme by coupling two tractors together, either physically in the shop or mechanically in the field.

However, there is a limit to how much traction a two-wheel-drive tractor can attain when pulling a load, even if it has the horsepower to do so. It worked for a while to use larger or wider tires, but with bigger tractors, that just led to more compaction. Manufacturers needed a way to put the power from bigger engines to the ground without damaging the fields. The answer came in the form of four-wheel-drive tractors that applied power to all four wheels. Equally important, both the front and rear tires on the traditional four-wheel-drive tractor are the same size to help spread the weight. With dual tires on each corner, compaction is decreased even more.

It's remarkable that some of the pioneers in development of four-wheel-drive tractors were working on similar ideas at the same time in different parts of the country. Moreover, the ideas didn't come from tractor manufacturers, but farmers and small companies that served the rural communities. Elmer Wagner and his brothers were already building concrete mixers and logging equipment in Oregon when they developed the first Wagner four-wheel-drive tractor. Although they had worked on the idea for several years, their final design was actually patented on September 24, 1953, with credit going to Elmer. The first model went on sale in 1955.

As you'll learn later in this chapter, the Steiger four-wheel-drive tractor came about in a similar manner. Brothers Doug and Maurice Steiger, who farmed in the famed Red River Valley, needed a larger tractor to cover more acres with bigger equipment. So they built the first Steiger tractor—mainly for themselves—in the family's dairy barn near Red Lake Falls, Minnesota. The machine impressed local farmers and neighbors to the point that the Steiger brothers were soon building four-wheel-drive tractors for others.

Andrew Morland

As sales took off for both companies, the major farm equipment companies soon took notice. However, not wanting to take the time they would need to engineer and develop their own model, many of them made deals with the existing companies. Hence, John Deere aligned with Wagner to market the John Deere WA-14 and WA-17 from 1968 to 1970 after experiencing problems with its own design. In the meantime, Allis-Chalmers, Ford, and International Harvester all contracted with Steiger at one time or another to build four-wheel-drive tractors for their respective brands while they completed designs of their own. At the same time, several other independent manufacturers were coming out with four-wheel-drive tractors. Among them were McConnell Tractors Ltd., which was owned by Ward McConnell, an entrepreneur who eventually acquired the rights to the Massey Ferguson four-wheel-drive tractor line in 1988. McConnell turned around and built four-wheel-drive tractors for Massey Ferguson from 1989 until 1993.

Other little-known four-wheel-drive models included the Rite tractors, created by Dave and John Curtis from Great Falls, Montana, and the Woods and Copeland tractors, which were designed by Texas engineer Jones Copeland and farmer J. D. Woods to handle the challenges inherent with rice farming. The brand was later sold in 1976 to the Rome Plow Company and moved to Georgia.

The Steiger and Wagner models weren't the first four-wheel-drive tractors on the market, though. Massey-Harris had introduced the General Purpose (GP) 15-22 nearly 30 years earlier with limited success. Even earlier, in 1915, John Fitch, a farmer from Mason County, Michigan, developed the Fitch Four-Drive tractor that used driveshafts and differentials on each axle to power the front and rear wheels.

The irony is that the Massey-Harris GP was supposed to be a general-purpose tractor that could be used for both row-crop work and tillage. Yet, the first four-wheel-drive models of the 1950s and 1960s were generally used for heavy-duty applications like ripping, plowing, and tillage. Today, four-wheel-drive tractors have come full circle to where they are now designed and used for everything from heavy tillage to planting to row-crop tillage. It just goes to show that some tractors, like some tractor features, were simply ahead of their time.

Massey-Harris General Purpose (GP) 15-22

Years of production	1930–1938
Engine	Hercules, 226 cubic inches (3.7 liters), 4 cylinders
Fuel	Gasoline
Horsepower	24.84 belt (tested), 19.91 drawbar (tested)
Rated RPM	1,200
Transmission	3 speeds forward, 1 reverse
Weight	3,900 pounds
Original price	$1,000

The Massey-Harris GP 15-22 was not only the first tractor designed by Massey-Harris, but one of the first tractors in the industry to feature a four-wheel-drive powertrain, when it came out in 1930. It was also Massey-Harris's first attempt at addressing the row-crop market with its 30 inches of under-frame clearance, and the first tractor to carry the Massey-Harris name only. Prior to the GP, the tractors had been sold under the Massey-Harris Wallis label following the purchase of the Wallis line in 1928.

The GP was offered in tread widths of 48, 56, 60, 66, and 76 inches to match various row spacings. However, the tread width wasn't adjustable once it had been ordered. In addition, it had right- and left-side braking, as well as a universal joint on the front axle, for short turns within a tight 6-foot radius—much like today's skid steer loaders.

The GP wasn't short on options, which were intended to add to its versatility. Among them were a PTO, implement lift, lights, orchard fenders, a starter, and extension controls that allowed the tractor to be controlled from the seat of a trailed implement. The four-wheel drive from equal-sized wheels also gave the GP traction that was matched by very few tractors.

Unfortunately, as a four-wheel-drive tractor, the GP was ahead of its time. Plus, its price tag of around $1,000 was much too expensive for the average farmer, who was now facing the Great Depression. It didn't help, either, that the GP had to compete with the less expensive, if not more popular, Farmall tractor—although the GP did find a following in the specialty-crop and forestry markets. In 1936, an "improved" version of the GP was introduced with a Massey-Harris engine of the same displacement. Unfortunately, it had less power than the original and was ultimately discontinued in 1938. After eight years of production, the GP disappeared from the market, and four-wheel-drive applications went back on the shelf for another 20 years.

The Massey-Harris GP was certainly ahead of its time in that it was one of the first four-wheel-drive tractors ever produced. However, it was also the first tractor actually designed and produced by Massey-Harris. *Ralph W. Sanders*

John Deere 8020

For all practical purposes, the John Deere 8020 four-wheel-drive tractor is nearly identical to the 8010 introduced just a year earlier in 1960. The 8010 was the first four-wheel-drive tractor offered by John Deere. It was also introduced at the same time and with the same styling as the New Generation tractors. However, due to transmission problems, all but one 8010, which supposedly escaped the recall, were converted to 8020 models.

As a result, it's been said that every 8020 in existence started out as an 8010 model. By the time the recall occurred, only a few of the 8010 models had been sold. So many of them were still on dealers' lots. Still, the 8010 was starting to earn a bad reputation, which prompted John Deere to load them all up on trucks and bring them back to the factory. Here, the tractors were fitted with a new transmission that included a new two-speed shifter on the floor. Each tractor was also equipped with a New Generation operator seat, along with new steps to the operator platform that were perforated, instead of solid deck plate. Finally, the tractors were repainted and fitted with new decals that designated them as 8020 models.

All totaled, only about 100 of the 8010 and 8020 models were built at the company's Waterloo, Iowa, plant. Of those, 49 were equipped with a three-point hitch, while the rest went out with a standard drawbar hitch. It was a full decade after the introduction of the 8010 that John Deere built another four-wheel-drive model in the 7020. In between, the company contracted with Wagner Tractor, Inc., to build the John Deere WA-14 and WA-17, which were based on current Wagner models.

Years of production	1961–1964
Engine	Detroit Diesel, 425 cubic inches (7.0 liters), 6 cylinders
Fuel	Diesel
Horsepower	215 engine; 150 drawbar (claimed)
Rated RPM	2,100
Transmission	8 speeds forward, 2 reverse
Weight	24,860 pounds
Original price	$30,000

Most John Deere 8020 four-wheel-drive tractors in existence are actually 8010 models that were recalled and retrofitted with a new transmission and some new features. All total, only 100 models were built before John Deere went outside the company for a four-wheel-drive model. *Andrew Morland*

Years of production	1966–1969
Engine	Case, 451 cubic inches (7.4 liters), 6 cylinders
Fuel	Diesel
Horsepower	119.90 PTO (tested), 106.20 drawbar (tested)
Rated RPM	2,000
Transmission	6 speeds forward, 6 reverse
Weight	16,500 pounds
Original price	$20,200 (1969)

Case 1200 Traction King

Case moved into the four-wheel-drive market in 1966 with the first of its Traction King models. Like all big four-wheel-drive models of the time, it featured the same large tires on the front and rear. However, unlike the other competitors on the market, the Traction King models did not have an articulated frame, which allowed pivot steering.

In contrast, the Case models actually featured four different modes of steering, which could be selected from the cab. These included conventional front-wheel-only steering or rear-wheel-only steering, the latter being particularly helpful for maneuvering the back of the tractor when hooking up an implement. For tight turns at the end of the field, four-wheel steering turned the front wheels and rear wheels in opposite directions, dramatically reducing the turning radius. The fourth steering mode was referred to as "crab steering," which turned all four wheels in the same direction. This was particularly valuable when performing tillage on side hills to overcome side drift by turning both sets of wheels slightly uphill. It could also be used in close quarters, such as maneuvering inside a machine shed or moving closer to a fence.

As for the machine itself, the 1200 was powered by a turbocharged version of the Case 451-cubic-inch diesel, first used in the 1030, to provide 120 engine horsepower. This was connected to a Clark R500 transmission that provided six speeds in each direction and forward speeds that ranged from 2.5 to 12.8 miles per hour.

Case didn't forget about operator comfort, either. Standard equipment included power steering, power brakes, a deluxe seat, and a convenient control console. Case was once again king of the wheatland tractors.

In 1969, the Traction King became part of the new 70 Series and was renamed the 1470. It was also upgraded with a new 504-cubic-inch turbocharged, direct-injected diesel that provided 145 horsepower, as well as hydrostatic steering on the front wheels and independent hydraulic steering on the rear.

The Case 1200 Traction King was unique in that it had a one-piece frame instead of one that was articulated. To compensate, both the front and rear wheels turned, providing a total of four different steering modes. *Ralph W. Sanders*

Versatile D100

V ersatile certainly wasn't the first to offer a modern-day four-wheel-drive model to the North American agricultural market when it introduced the Model G100 and D100 in 1966. International Harvester, John Deere, Steiger, and Case had all introduced four-wheel-drive machines to the market with limited success—nobody was selling more than a few hundred models per year.

Peter Pakosh and Roy Robinson, co-owners of Versatile Manufacturing Ltd., based in Winnipeg, Manitoba, were determined to take a different route. They felt that four-wheel-drive tractors had a place in the industry, if only they were priced closer to what the average farmer could afford. So they

Years of production	1966–1967
Engine	Chrysler, 318 cubic inches (5.2 liters), 8 cylinders
Fuel	Gasoline
Horsepower	100 drawbar (claimed)
Rated RPM	2,400
Transmission	12 speeds forward, 4 reverse
Weight	Unknown
Original price	$7,500

set out from the start with a goal of building and marketing a four-wheel-drive tractor, with all its power and traction benefits, for around the same price as a two-wheel-drive tractor from one of the major brands.

Although the partners didn't have any experience building tractors, they did have a large share of the North American market for sprayers, grain augers, and swathers. All were fairly basic machines that were built to do the job, built to give good service, and built to sell at a price the average farmer could afford.

In order to keep the price low, Versatile D100 and G100 were mass-produced using readily available components. The designers started with a solid, articulated frame built in their own Winnipeg factory. While the D100 used a 363-cubic-inch Ford inline six-cylinder diesel engine, the G100 featured a 318-cubic-inch, gasoline-powered Chrysler V-8 engine. Attached to that was a Spicer transmission and Funk transfer case, which, together, provided 12 speeds forward and 4 speeds in reverse. The drivelines were also built by Spicer, while the front and rear axles were from Clark and the hydraulic pump was an off-the-shelf Cessna model.

Thanks to the simplicity and commonality of parts, the Versatile D100 was introduced with a price of $9,200. The gasoline-powered G100 was even cheaper at $7,500. In contrast, a John Deere 8010 had a base price of nearly $30,000, and a Case 1200 was approximately $20,200—although both were a lot more complicated, which obviously contributed to their price.

One year later, in 1967, Versatile introduced its new G125 gas-powered and D118 diesel-powered models. Like the G100 and D100, the new models were well built with reliable components, but were manufactured without frills and options to keep the price within reach of the average farmer. As a result of the low price and simplicity of the 100 models and the 2 models that followed, Versatile was selling more four-wheel-drive models by its second year of production than any other manufacturer in the industry.

Although Versatile wasn't the first to offer a four-wheel-drive tractor when it introduced the G100 and D100, they were the first to produce a model that was priced competitively with a two-wheel-drive tractor. Thanks to the no-frills approach, Versatile tractors soon began outselling all other models combined. *Don Wadge*

Minneapolis-Moline A4T 1600 Diesel

Like every other tractor manufacturer in business at the time, Minneapolis-Moline dealers saw a market for a four-wheel-drive tractor and wanted to catch a piece of the action. Of course by this time, Oliver, Minneapolis-Moline, and Cockshutt had been merged together in 1969 to form White Farm Equipment. Since the three brands were still being marketed separately in their respective colors, the first four-wheel-drive models released by White were introduced under three brand names and two model numbers.

Years of production	1970–1974
Engine	Caterpillar, 636 cubic inches (10.4 liters), 8 cylinders
Fuel	Diesel
Horsepower	225 engine, 159 drawbar (claimed)
Rated RPM	1,200
Transmission	10 speeds forward, 2 reverse
Weight	19,600 pounds
Original price	Unknown

In Minneapolis-Moline's case, it was the A4T 1400 and A4T 1600. The A4T 1400 was the first to hit the market in 1969, followed by LPG and diesel versions of an A4T 1600 a year later. Built in Hopkins, Minnesota, in a plant owned by White, and later in Charles City, Iowa, the Minneapolis-Moline A4T 1600 was also sold as the Oliver 2655 and as the White Plainsman A4T-1600. The Plainsman was sold only in Canada, since White had started to displace the Cockshutt name with the Plainsman name.

The LPG version of the Minneapolis-Moline A4T 1600 used a 504-cubic-inch engine rated at 2,200 rpm and 169 engine horsepower. The diesel version, meanwhile, had a Minneapolis-Moline 585-cubic-inch engine rated at 169 horsepower.

Oliver, Minneapolis-Moline, and Cockshutt had already been merged together to form White Farm Equipment by the time their first four-wheel-drive tractor was introduced. Yet, the brands were still marketed separately, which allowed the larger of the company's two new models to be marketed as both the Minneapolis-Moline A4T 1600, Oliver 2655, and White Plainsman A4T 1600. *Andrew Morland*

Both models used a standard 10-speed transmission that offered speeds from 2.0 to 22.2 miles per hour. The articulated frame offered 44 degrees of oscillation in each direction for tight turns at the headland.

In 1974, White consolidated the Oliver, Minneapolis-Moline, and Plainsman into a newly designed 150-horsepower White four-wheel-drive model. Introduced as the White 4-150 Field Boss, it incorporated the new Field Boss styling and silver and charcoal gray color scheme—marking the final demise of Oliver green, Minneapolis-Moline yellow, and Cockshutt red in large tractors.

Ford FW-60

Ford had one of the best reputations in the business for small and midsize tractors when they decided to expand into the high-horsepower, four-wheel-drive market in 1976. Unfortunately, the company had never built a high-horsepower tractor, let alone a four-wheel-drive model, before. That is why it turned to Steiger, just as other companies before it had done.

Under the agreement, Steiger actually supplied four different tractors as part of the Ford FW Series. All were painted in blue and white and labeled with the Ford name and FW model number. All were also powered by Cummins V-8 diesel engines. At the bottom of the range was the FW-20 at 210 engine horsepower, while the top spot was occupied by the FW-60 with 335 horsepower from a turbocharged Cummins.

A 20-speed transmission literally provided a gear for every application, including transport at up to a brisk 21.8 miles per hour. To keep the operator comfortable, the FW cab was equipped with tinted glass, standard-equipment air conditioning, a stereo sound system that included a radio and an 8-track tape player, and soundproofing insulation.

In 1980, Ford upgraded the FW-60 with some minor styling and engine changes and a year later added the option of a 10-speed automatic transmission. The supply chain changed altogether, though, in 1986. That's when Case IH acquired the Steiger Company through a successful takeover

Years of production	1976–1982
Engine	Cummins, 903 cubic inches (14.8 liters), 8 cylinders turbo
Fuel	Diesel
Horsepower	335 engine, 270.87 drawbar (tested)
Rated RPM	2,600
Transmission	20 speeds forward, 4 reverse
Weight	31,100 to 33,970 pounds
Original price	$99,394 (1982)

Built by Steiger, the FW Series tractors helped establish Ford in the four-wheel-drive market. As part of the agreement, Steiger built four different FW models, including the FW-60 shown here. However, since the original models were blue and white, this particular model has obviously been repainted. Ironically, Ford purchased Steiger shortly after the FW line was discontinued. *Andrew Morland*

bid. A year later, Ford bought the Canada-based Versatile tractor company to secure its own supply of four-wheel-drive tractors.

Big Bud 16V-747

There was only one model of the Big Bud 16V-747 tractor ever built by Ron Harmon and the crew of the Northern Manufacturing Company in Havre, Montana. There had been other Big Bud tractors and models built, but nothing had ever been this big. Designed to produce a whopping 760 horsepower, the unusual tractor measured 27 feet long, 20 feet wide, and 14 feet tall. In fact, it was so large, the tires had to be specially made by United Tire Company in Canada in order to meet the desired 8-foot diameter. That's part of the reason the tractor never actually went into production after the first one was built. It was simply too large to move.

The tractor was originally designed for and purchased by the Rossi Brothers, who were cotton farmers near Bakersfield, California. They owned and used the 16V-747 for 11 years, mostly for deep ripping fields. At that point, the tractor was moved all the way across the country to Indiantic, Florida, where Willowbrook Farms used it for deep ripping, as well.

It wasn't until 1997 that the 16V-747 was returned to Montana—only 60 miles from where it was built—and put to work by the Williams Brothers, who farmed near Big Sandy. There, the tractor was used for field cultivation, covering up to one acre per minute with an 80-foot cultivator. In the meantime, the Williams Brothers gave the tractor a new paint job, new chrome mufflers, and a power boost that took the tractor to 900 engine horsepower.

Years of production	1977
Engine	Detroit Diesel, 1,472 cubic inches (24.1 liters), 16 cylinders
Fuel	Diesel
Horsepower	900 engine
Rated RPM	2,100
Transmission	6 speeds forward, 1 reverse
Weight	100,000 pounds
Original price	$300,000

The only model of its kind, the Big Bud 16V-747 actually started its life as a 760-horsepower model. However, in 1997, after being in use for 20 years, it was refurbished and modified to produce a whopping 900 horsepower. *Ralph W. Sanders*

More recently the tractor has been making the tractor show circuit and has been housed at the Heartland Museum in Clarion, Iowa, in between special appearances.

Allis-Chalmers 4W-305

Years of production	1982–1985
Engine	Allis-Chalmers, 731 cubic inches (12.0 liters), 6 cylinders
Fuel	Diesel
Horsepower	305 engine, 254 drawbar (claimed)
Rated RPM	2,400
Transmission	20 speeds forward, 4 reverse
Weight	27,100 pounds
Original price	$93,685

The Allis-Chalmers 4W-220 and the bigger 4W-305 four-wheel-drive tractors were first shown to Allis-Chalmers dealers at a new product introduction in Reno, Nevada, in January 1982. They were accompanied by four new 8000 Series tractor models, new 6000 Series cabs, and a new 40-horsepower 6140. With the 6000s being less than 18 months old, Allis-Chalmers essentially had an all-new line from 40- to 250-PTO horsepower for the first time in Allis-Chalmers history.

Compared to the 8550 it replaced, the 4W-305 was actually very similar in horsepower. While the 8550 was rated at 300 engine horsepower and 250 PTO horsepower, the 4W-305 was rated at 305 engine horsepower and 254 (claimed) PTO horsepower. The 6120T engine built by Allis-Chalmers featured a 5.250x5.625-inch bore and stroke. However, the rated speed for the Allis-Chalmers 731-cubic-inch (12-liter), 6-cylinder turbocharged diesel engine used in both the 4W-305 and its predecessor was reduced 150 rpms in the 4W-305. The reduction from 2,550 to 2,400 rpm was offset by an increase in torque to maintain the same horsepower rating. It also helped control engine noise and vibration, while a new injection system improved performance and reduced smoke.

The biggest change, though, involved the cab, sheet metal, and amenities. As the "widest cab on a four-wheel-drive tractor," it incorporated all the improvements made in the 8000 Series, including a new seat with better suspension and more adjustment, an improved climate-control system, more convenient controls, and a "well-appointed interior." For added comfort, the cab also incorporated large rubber cab mountings, external air intake filters, and heavy insulation for even better soundproofing.

Rated at 305 engine horsepower, the 4W-305 was not only the most powerful tractor Allis-Chalmers had ever built, but it was unfortunately the last four-wheel-drive tractor the company built before it was acquired by Klöckner-Humboldt-Deutz (KKHD) in 1985. *Andrew Morland*

Like the 8000 cab, the 4W-305 was designed like a truck cab, in that it had a solid firewall and cowl that were part of the cab, not part of the tractor. The instrument cowl, all controls, and the seat were preassembled in the cab, then it was mounted on the tractor in a body-drop operation.

Between the time the 4W-305 was introduced and when the company ceased production upon the purchase of Allis-Chalmers by Klöckner-Humboldt-Deutz in 1985, the company built 412 copies of the 4W-305. It was an unfortunate, sudden end for arguably the best four-wheel-drive tractor Allis-Chalmers ever built.

Years of production	2000–2001
Engine	Cummins, 912 cubic inches (14.9 liters), 6 cylinders turbo
Fuel	Diesel
Horsepower	440 engine, 345.68 drawbar (tested)
Rated RPM	2,000
Transmission	16 speeds forward, 2 reverse
Weight	51,335 pounds
Original price	Unknown

Case IH STX440

One of the first and largest four-wheel-drive tractors to come out of the former Steiger factory in Fargo, North Dakota, following the purchase of the company by Case IH was a new STX440. Painted Case IH red, instead of the former Steiger green, the tractor was part of a new series that ranged from 275 to 440 horsepower. Being at the top of the range, the STX440 was, at least for a while, the most powerful commercially available tractor.

Equipped with a 904.4-cubic-inch Cummins diesel, it featured a 43 percent power reserve and attained its 440-horsepower power rating at only 2,000 rpm. The low rpm was said to improve fuel economy, while reducing both vibration and noise levels.

The STX440 was unusual in yet another way, too. It was available with the standard rubber tires—usually duals on all four corners—just like any other four-wheel-drive tractor. However, it was also available with the new Case IH Quadtrac (QT) system, which used a pair of separate track drives in place of the wheels on each axle. This allowed the tractor to articulate, just as it would on tires, yet provide the flotation of tracks.

Another unique feature was found on the STX 16-speed powershift transmission. Thanks to electronic shifting, Case IH was able to incorporate a feature called Autoskip. This allowed the transmission to skip every other gear when accelerating from a stop. The benefit was it took much less time to reach a travel speed than going through each gear at a time.

Challenger MT975B

With the introduction of the 570-horsepower MT975B Series four-wheel-drive tractor in 2006, Challenger set a new record for the industry's largest production four-wheel-drive articulated tractor. The key word in that statement was "production," since larger tractors, including the Big Bud 16V-

The STX440 was one of the first and largest four-wheel-drive tractors to come out of the former Steiger factory in Fargo, North Dakota, following the purchase of the company by Case IH. Even more unique was the fact that it was the first four-wheel-drive tractor to offer the Quadtrac system, which used a set of tracks in place of tires on each axle. *Ralph W. Sanders*

747 and 750-horsepower Rite "Earthquake," had already been built. They just hadn't been built in commercial numbers.

The MT975B was part of the new MT900B Series that included four model sizes from 430 to 570 engine horsepower—the MT975B being the largest. However, all of the Challenger MT900B Series tractors shared nearly 80 percent of their components with the company's Challenger track tractors. Common features included a Caterpillar ACERT engine, a Caterpillar-built 16FX4R powershift transmission, Challenger's computerized Tractor Management Center, and a unique Load Independent Flow Division hydraulic system that provided a standard flow rate of 43.5 gallons per minute with an option of 59 gallons per minute. Challenger's largest track and wheeled models also shared a 108-cubic-foot cab that included such amenities as an

Years of production	2006–2009
Engine	Caterpillar, 1106 cubic inches (18.1 liters), 6 cylinders
Fuel	Diesel
Horsepower	570 engine, 19.91 drawbar (tested)
Rated RPM	1,800
Transmission	16 speeds forward, 4 reverse
Weight	60,000 pounds
Original price	Unknown

Patterned after the Challenger MT800 Series track tractors, the MT900B Series four-wheel-drive tractors shared nearly 80 percent of their components with their tracked counterparts. The first four-wheel-drive tractors to carry AGCO Corporation's Challenger brand name, the MT975B also set a new record for horsepower among production four-wheel-drive tractors. *Tharran Gaines*

air-suspension seat, a standard four-speaker sound system, and a Surround-Flow ventilation system.

Just a year earlier, Challenger had introduced the second-generation MT875B track tractor at 570 horsepower, which set the benchmark for power in an agricultural setting and broke all existing records for drawbar horsepower at the Nebraska Tractor Tests. With the release of the MT975B, after years of testing, Challenger had a four-wheel-drive tractor with an equal amount of horsepower. Of course, by this time Challenger was part of AGCO Corporation, having been purchased from Caterpillar in 2002.

It would be a mistake, though, to think that the MT900B Series tractors were simply track tractors on wheels. The 120-millimeter bar axles, for example, were the largest standard axles in the industry at the time. The articulation joint was also held together with the largest pin in the industry. Yet, it offered a 42-degree turning radius and 13 degrees of oscillation.

Today, the Challenger MT900 Series tractors continue to break ground throughout the world. In fact, in 2013, AGCO launched the MT900E Series tractors with even more features, technology, and horsepower. As the most recent version, the MT975E boasts 590 engine horsepower and 425 PTO horsepower.

Record Sales

As you look through the next few pages, it shouldn't come as a big surprise that most of the best-selling tractors of all time were built and introduced in the 1930s and 1940s. One of the biggest reasons was that over half of the nation's farmers were still using horses at that time.

In 1930, farmers still accounted for 21 percent of the labor force, and amazingly, there were 6,295,000 farms in America, with an average size of 157 acres. However, most tractors offered at that time were bigger models that were either used by large farms or by custom operators, such as those who went from farm to farm, threshing wheat with a threshing machine, shelling corn, or perhaps baling hay with a stationary baler. More than half of the farms were still smaller than 100 acres, which meant they couldn't justify a tractor just for tillage or large equipment.

That situation was all about to change with the development of the row-crop tractors. Naturally the larger operations were the first to adopt the versatile new machines that could do plowing and cultivating, as well as everything in between. Those farmers accounted for the tremendous sales volume of tractors like the John Deere A and Farmall F-20. Most machinery companies realized, however, that there was an equally large market for a smaller tractor that those with fewer than 100 acres could afford. Those more compact models accounted for much of the second wave of huge sales.

Several tractor manufacturers even tried to do the math for their audience as part of their marketing campaigns. The late Norm Swinford, author of *Allis-Chalmers Farm Equipment 1914–1985* and a member of Allis-Chalmers' marketing department for 30 years, recalled, "Advertising and sales promotion concentrated on the

Ralph W. Sanders

high cost of horse farming, emphasizing that it took five acres of cropland to feed one horse; 25 acres for a five-horse operation. 'Being chambermaid to five horses takes 270 hours per year—why not use that time and the 25 acres to raise profit-producing livestock?'" he added, quoting the ad copy. "The Models B and RC, and later the Model C, effectively retired ol' Dobbin to the bluegrass pasture."

As we saw back in Chapter 1: First in the Field, even Henry Ford had the cost of owning a team of horses or mules in mind when he challenged the engineering team to "build a tractor that would cost no more than the combined cost of a team of horses or mules, the harness and the 10 acres of land required to produce enough feed for the team."

Of course, Henry Ford was also a master of mass production and assembly line manufacturing, which helped introduce both the Fordson tractors and the Ford N Series tractors that followed a few years later at a price that a growing number of farmers could afford.

The other thing that accounted for such high sales volume for a number of models was the length of time they were on the market. Even though the model might have received a larger engine or different transmission along the way, the model number remained unchanged. The John Deere D, for example, was on the market as the "D" for 30 years, while others remained as part for the inventory for 10 to 20 years. Most models were also available in a number of different versions.

Today, tractor manufacturers are generally so determined to stay ahead of the competition that tractors are assigned new model numbers or series designation any time a noteworthy change or redesign occurs. Plus, each series generally encompasses three to five models to provide the customer with a model that is just the right size.

Of course the biggest reason you will never see a single tractor model boast sales of a half-million units, or even 200,000, is because the market no longer supports such volume. Farmers aren't buying tractors for the first time to replace horses or hand labor. They're simply buying replacement tractors or adding another tractor to handle a particular job. Today's tractors are also doing the job that two, three, or even four tractors did in the 1930s. So fewer tractors are needed, even if the total acreage in farmland had remained the same. Instead, it's been reduced over the years due to urbanization and population growth—8 percent in the past 20 years alone.

Finally, the number of farms and farmers is lower than it's ever been in history. The number of farms in the United States actually peaked in 1935 at 6.8 million. Notice again how that date coincides with peak tractor sales. By 2007, the total number of farms was down to 2.2 million. Yet, many of those were small farms, hobby farms, and so forth. Of that total, approximately 187,815 accounted for 63 percent of the sales of agricultural products.

It's no wonder so many tractor companies have merged over the years and so many other brands have simply disappeared into history.

Fordson F

Having grown up on a farm, Henry Ford had an interest in tractors that rivaled that of automobiles. In fact, in 1906, before he ever became famous for his Model T, Henry Ford built what he called his Automobile Plow. It was propelled by a four-cylinder engine and transmission taken from his Model B touring car. However, by 1917, Henry Ford was ready to enter the tractor market in earnest with a new brand and a new design. Built by a separate division called Henry Ford and Son, the tractor was simply named the Fordson. There was at least one, and possibly two, reasons Ford settled on the Fordson name, rather than the name previously made famous on his automobiles. For one, the name Ford was already being used on a tractor built in Minneapolis, Minnesota. Not only was the Minneapolis Ford—which obviously tried to capitalize on the Ford name—one of the worst tractors ever built, but if the company had visions of Henry Ford paying top dollar for the Ford tractor name, that wasn't going to work either. Some theorize that Ford thought the Fordson name was best from the beginning since some Ford Motor Company shareholders didn't approve of tractor production. At any rate, the Fordson F was introduced in 1917, with a 20-horsepower four-cylinder engine built by the Hercules Engine Company in Ohio. (It would

Years of production	1917–1928
Engine	Hercules, 251 cubic inches (4.1 liters), 4 cylinders (until 1920)
	Ford, 251.3 cubic inches (4.1 liters), 4 cylinders (1920 and later)
Fuel	Distillate
Horsepower	22.28 belt (tested), 12.32 drawbar (tested)
Rated RPM	1,000
Transmission	3 speeds forward, 1 reverse
Weight	2,710 pounds
Original price	$395 (1922)

Built from 1917 to 1928, the Fordson Model F became one of the best-selling tractors in history, partly due to its affordable price but also because it was sold worldwide with approximately 6,000 units seeing duty in Great Britain during World War I. *Ralph W. Sanders*

switch to a similar Ford engine in 1920.) Founded in 1915, Hercules Engine Company was known for its strong industrial-type heavy-duty engines designed for the rapidly expanding trucking industry.

Not only was the Fordson F smaller than most other tractors, which made it more affordable and easier to produce, but it also lacked a conventional frame. Instead, the engine, transmission, and axle housings were all bolted together to form the basic structure of the tractor. That practice started a trend that would carry on through the N Series Ford tractors and most of Ford tractor history. The unique design and shorter dimensions of the F turned out to be both good news and bad news. The advantage was that it was short enough to set crossways on a railcar so that more of them could be shipped at once at a lower cost. The bad news was that because of its light weight and short wheelbase, the F was subject to flipping over backward if a load was improperly applied.

The Fordson market took a turn though during the Great Depression, when sales began to fall in the United States for all tractors. Consequently, Henry Ford ceased production in the United States and transferred Fordson production to Ireland in 1928.

Due to the high cost of shipping Fordson farm tractors to the United States from Britain, not to mention competition from a growing number of American companies, Fordson's market share among American consumers slipped to a low of only five percent even though a new, more powerful N had been introduced. Still, the Fordson F became one of the largest-selling tractor models in the world, with more than 700,000 F tractors being produced in Dearborn, Michigan, before production was moved to Great Britain, where another 7,597 tractors were built. Even before production left the United States, Fordson F tractors were being shipped around the world. The British government, for example, bought 6,000 of the tractors for use during World War I. Russia also imported more than 26,000 Fordson tractors in an effort to increase food production, while building an unknown number of Fordsons themselves under license from Ford.

While the Fordson F was a well-built versatile tractor on its own, Henry Ford's price-cutting policies also had a lot to do F sales. In 1918, the F was listed at $750. However, that price was reduced several times to the point that a Model F in 1922 cost a farmer only $395.

McCormick-Deering 10-20

The McCormick-Deering name was first used on the International Harvester 10-20 and 15-30 tractors to designate tractors with a standard or wide front axle, while the Farmall name was originally reserved for row-crop tractors. This nomenclature continued for nearly three decades until the two names were eventually phased out in favor of the International Harvester name alone. Of course, the McCormick part of the name goes back to 1831, when Cyrus McCormick produced the first successful reaper.

Introduced shortly after the more powerful 15-30, the 10-20 shared similar styling and was mechanically similar except for the smaller size and less-

Years of production	1923–1939
Engine	International Harvester, 283.7 cubic inches (4.6 liters), 4 cylinders
Fuel	International Harvester, 283.7 cubic inches (4.6 liters), 4 cylinders
Horsepower	24.8 belt (tested), 19.60 drawbar (tested)
Rated RPM	1,000
Transmission	3 speeds forward, 1 reverse
Weight	4,010 pounds
Original price	Unknown

powerful engine. While the 10-20 featured a 283.7-cubic-inch engine, the 15-30 was fitted with a 381.7-cubic-inch engine rated at 10 more horsepower. Both, however, featured a three-speed gearbox that provided top speed of 4 miles per hour.

After four years, the 10-20 was given a design update that increased the rated engine speed to 1,025 rpm and a slightly faster top speed of 4.25 miles per hour. Rubber tires were added as an option in the late 1930s.

One of the most unique features of the 10-20 was the fact that it had a PTO as standard equipment, making it one of the first in the industry to be so equipped.

McCormick-Deering already had a familiar name, thanks to the McCormick reaper, when it introduced the 10-20 tractor in 1923. Adding to its appeal was the fact that it was one of the first in the industry to feature a standard-equipment PTO. As a result, more than 215,000 models were sold by the time it was replaced in 1939. *Ralph W. Sanders*

Available as an orchard model, as well as the standard-tread model, the 10-20 saw annual production peak at 34,742 in 1929, with a sales total of 215,000 units by the time the tractor was replaced in 1939.

John Deere D

As the first totally new tractor designed by John Deere, the D carried on much of the tradition of the Waterloo Boy from which it was derived, including the two-cylinder engine and a number of shared components. Like its predecessor, it was also considered to be a wheatland tractor, which meant it was designed for pulling plows and providing the power for threshers. However, that was the typical role of most farm tractors at that time, having picked up the jobs from steam tractors.

In addition to being the first completely new tractor offered by John Deere, the D was also the longest-running production model in the John Deere stable. Over its 30-year lifespan, which ran from 1923 to 1953, some 160,000 units were built and sold. Yet, there were literally dozens of variations within the D line. In addition, there were numerous production changes in horsepower, features, and style that set them apart.

The first 50 Model D tractors, for instance, had a 26-inch flywheel, as well as several features that were not found on models that followed. Nevertheless, its 26-inch flywheel and a one-piece steering shaft continued from serial number 30451 through 31279. Beginning with serial number 31280 and continuing through serial number 36248, a 24-inch spoke flywheel was used, along with a two-piece steering shaft. Finally, in 1925 with serial number 36249, a solid flywheel was adopted. Today, two-cylinder tractor enthusiasts still refer to early Ds as spokers, which naturally makes those models extremely collectable.

Model D collectors can also tell you about the dozens of other variations that occurred over the years, making each one unique. That includes differences in fender width, exhaust stack type, front axle configuration, wheel type, and so on.

Years of production	1923–1953
Engine	John Deere, 465 cubic inches (7.6 liters), 2 cylinders (1923–1927)
	John Deere, 501 cubic inches (8.2 liters), 2 cylinders (1927–1953)
Fuel	Distillate
Horsepower	41.59 belt (tested), 29.90 drawbar (tested) (1953)
Rated RPM	800 (900 on 501-cubic-inch-displacement engine)
Transmission	2 speeds forward, 1 reverse (3 speeds forward after 1935)
Weight	4,090 to 5,880 pounds
Original price	$1,000 (1923) to $2,124 (1953)

Although it saw several revisions, modifications and horsepower increases over its 30-year run, the D holds the longest-running production run of any tractor in the entire John Deere lineup. Because it was sold as the same model for three decades, total sales volume was predictably high at over 160,000 units. *Ralph W. Sanders*

Between the first model and the last, customers also saw the belt horsepower go from just over 27 to nearly 42 horsepower by 1953. In 1935, John Deere added a third transmission speed. Perhaps the biggest change, though, came in 1939, when John Deere released the styled tractors that featured bodywork designed by renowned industrial designer Henry Dreyfuss.

Early D models, known affectionately as "spokers" due to their spoked flywheel, are still sought after by John Deere collectors. That's particularly true of the first 50 models, which feature a 26-inch flywheel in place of the 24-inch spoked flywheel that followed. *Ralph W. Sanders*

Farmall F-20

Years of production	1932–1939
Engine	International Harvester, 220.9 cubic inches (3.6 liters), 4 cylinders
Fuel	Distillate or gasoline
Horsepower	28 belt (tested), 20.66 drawbar (tested)
Rated RPM	1,200
Transmission	4 speeds forward, 1 reverse
Weight	4,400 pounds
Original price	$895

Given the success and popularity of the Farmall Regular, which was essentially the industry's first row-crop tractor, it didn't take long for other manufacturers to develop their own narrow-front row-crop tractors, including John Deere, which introduced its GP (general-purpose) model in 1929. In response, International Harvester introduced the Farmall F-20—an improved version of the Farmall Regular. Although it used the same 220-cubic-inch engine, improvements in the head and piston design provided a greater power level of 28 horsepower on the PTO/belt and 20.66 horsepower on the drawbar, according to the Nebraska Tractor Test

As an improved version of the Farmall Regular, which started the whole general-purpose or all-purpose tractor category, the subsequent Farmall F-20 saw even more demand as farmers turned to mechanization in order to become more efficient. As a result, just over 154,000 F-20 tractors were sold in just nine years. *Ralph W. Sanders*

Laboratory. The F-20 was also heavier than the Regular at 4,500 pounds and slightly longer. Plus, it came with a four-speed transmission in place of the three-speed, as well as new options, including a wide front end, a narrow rear tread, and eventually, rubber tires. With prices ranging from $895 to $1,000, it became International Harverster's most popular tractor until production ended in 1939. During the nine years the F-20 was sent to dealers, just over 154,000 models were sold.

Farmall F-12

Years of production	1932–1938
Engine	International Harvester, 283.7 cubic inches (4.6 liters), 4 cylinders
Fuel	Gasoline or distillate
Horsepower	16.20 belt (tested), 12.31 drawbar (tested)
Rated RPM	1,400
Transmission	3 speeds forward, 1 reverse
Weight	2,700 pounds
Original price	$525

After introducing the F-20 and F-30, following the success of the Farmall Regular, International Harvester discovered that these two newest models were still too large for some farmers, particularly those who continued to farm with horses. The answer came in 1932 in the form of the McCormick-Deering Farmall F-12. At around $525, it cost about one-third less than the F-20 and provided enough power for a one-bottom plow, a two-row planter, or a cultivator or sickle-bar mower. In fact, a full line of "quick-detachable" implements could be "on and off in a jiffy." Available with either a gasoline or distillate fuel system, the F-12 produced a maximum of 16.20 horsepower on the belt/PTO and 12.31 on the drawbar. Although the first F-12 models were equipped with a Waukesha engine, International Harvester soon replaced it with a four-cylinder engine of its own design that featured overhead-valves and a 113-cubic-inch displacement.

Unlike the previous Farmall models, though, the F-12 featured a unique stub-frame/transmission design and larger-diameter wheels that provided the desired crop clearance without the need for bull-gear drop boxes at the ends of the axle. Both rubber tires and a wide front end were optional.

Finally, in 1938, the F-12 was replaced by the F-14. In many respects, it was the same tractor except for a little more power and a raised steering wheel that provided more operator comfort. Due to a faster governor setting, the same 113-cubic-inch engine used in the F-12 now generated 17 maximum belt horsepower and 13.24 on the drawbar. Production of the F-14 only lasted

At around $525, the Farmall F-12 cost about one-third less than the F-20 and provided enough power for a one-bottom plow, a two-row planter, or a cultivator and/or sickle-bar mower. That made it ideal for the farmer who couldn't afford a larger tractor and was still using horses for certain jobs. Those facts alone created the demand for more than 120,000 models. *Ralph W. Sanders*

two years, though. By 1939 the F-14, along with the unchanged F-20 and F-30, would be replaced by a new generation of Farmalls.

During its production run, which lasted from 1932 through 1938, more than 120,000 F-12s were sold. However, that includes Fairway versions, which were interspersed in the production.

Allis-Chalmers WC

The Allis-Chalmers WC was one of those tractors that turned out to be the right machine at the right time. Allis-Chalmers already had a strong reputation by this time, and the purchase of Advance-Rumely in 1931 had provided the company with a broad network of dealers. Plus, the innovation

Thanks to a broad network of dealers acquired through the purchase of Advance-Rumely and a totally new tractor designed for using rubber tires, the Allis-Chalmers WC saw sales that totaled over 178,000 between unstyled and styled models. *Ralph W. Sanders*

of rubber tires, which had previously been introduced on the Model U, was still on the minds of farmers everywhere.

Launched in 1933, the WC was marketed as a two-plow tractor poised to take on the likes of John Deere and Farmall. Although only 3,000 WC models were sold the first year, sales soon took off, totaling over 10,000 by the second year and 29,000 in 1937, which was the peak year for sales. Even when it was discontinued in 1948, the WC was still selling well, leading to a total sales tally of 178,202 units.

One of the tractor's main selling points was its high power-to-weight ratio. Although a Waukesha engine powered the first 25 WC tractors, Allis-Chalmers began using its own 201-cubic-inch engine in early 1934. The Allis-Chalmers engine was a bit unique in that it featured what was called a "square engine," which meant the stroke and bore were identical at 4 inches each, providing a rated 21 horsepower at the drawbar. Yet because the engine was lighter and the chassis was made out of lighter-weight, high-tensile steel, the WC was able to perform as well as some more powerful tractors.

The WC also had the earned reputation of rubber tires behind it, as Allis-Chalmers had been the first to use them on a farm tractor. In fact, the WC was actually designed for rubber tires. By now, the skepticism toward them had declined, especially after the Nebraska Tractor Test Laboratory showed that rubber tires were up to 25 percent more fuel efficient, provided better traction, were easier to steer, allowed higher speeds on the road, and were quieter to operate. It's little wonder most buyers opted for rubber tires, even though they added $150 to the price of the tractor, compared to steel wheels.

Like its competitors, the WC was available in different versions, including a cane model and a specialty-crop model with a single front wheel. Like John Deere and Farmall, it also took on new styling during its tenure, gaining a new look in 1938 when it was restyled to match the B.

Years of production	1933–1948
Engine	Waukesha, 186 cubic inches (3.0 liters), 4 cylinders (on first 25 built) Allis-Chalmers, 201 cubic inches (4.6 liters), 4 cylinders (SN 7405 and later)
Fuel	Gasoline or distillate
Horsepower	29.93 belt (tested), 22.29 drawbar (tested)
Rated RPM	1,300
Transmission	4 speeds forward, 1 reverse
Weight	3,190 to 4,545 pounds
Original price	$825 (1934 WC with rubber tires)

John Deere A

With the introduction of row-crop, general-purpose tractors by some of its toughest competitors, Deere couldn't ignore its need for a more versatile model. The company's first successful attempt came when it released the GP in 1928. However, nothing sparked the sales of John Deere two-cylinder tractors like the introduction of Deere's second-generation row-crop tractors. This newest series began with the A in 1934. For once, John Deere had a model that could compete head-to-head with International Harvester's Farmall. Equipped with a tricycle row-crop front end, the A also featured an adjustable rear-wheel tread width, a hydraulic power lift, and individual right and left foot brakes that were geared directly to the large-diameter drive gears.

The A was rated for two 16-inch plows and produced 18.7 drawbar (24.7 belt) horsepower, thanks to its 5½ x 6½-inch cylinder distillate engine.

Years of production	1934–1952
Engine	John Deere, 309 cubic inches (5.1 liters), 2 cylinders (until 1940) John Deere, 321 cubic inches (5.3 liters), 2 cylinders (1940 and later)
Fuel	All fuel
Horsepower	24 belt (claimed), 16 drawbar (claimed)
Rated RPM	975
Transmission	4 speeds forward, 1 reverse (6 speeds after 1940)
Weight	3,525 to 4,909 pounds
Original price	$2,400 (1952)

The John Deere A was the first model released as part of the company's second generation of row-crop tractors. Designed as a two-plow tractor, it had the pulling power of a team of 6 horses and a daily work output that exceeded 10 horses. It didn't take much math to decide a new A was the cheaper investment. *Ralph W. Sanders*

Company literature at the time claimed that the A had the pulling power of a team of 6 horses and that it had a daily work output that exceeded that of 10 horses. In 1939, a generator and lights became available as an option on the A, making a tractor even more valuable since a farmer could work at night—something he couldn't do with horses.

Between 1934 and 1952, just over 300,000 units of the A were built in various configurations in both styled and unstyled versions, including the AR standard-tread model and AO orchard model. Along the way, John Deere also increased the horsepower twice. The first time was in 1939, when a longer stroke and bigger valves bumped the horsepower to 26 on the belt and 20 on the drawbar. The second time was in 1947, when a higher compression ratio of 5.6:1 increased the horsepower to 33 on the belt and 26 on the drawbar.

John Deere B

Despite the growing use of tractors in the 1930s, there were still some people who thought the John Deere A was too large. So Deere answered the demand with the B. Promoted as being two-thirds the size of the A, the B weighed about 780 pounds less than the A and utilized two 4¼ x 5¼-inch pistons to generate 11.8 drawbar/16 belt horsepower. Later versions, though, featured more than double the drawbar horsepower. Whether it was because of its smaller size, lower price, or greater versatility, the B seemed to be the perfect choice for a small farm. With production runs totaling 322,200 units, it

Like many companies in the 1930s, John Deere no sooner had one market covered than it moved on to the next, which called for either a larger model or a smaller one. The John Deere B covered the latter. With only 16 horsepower on the belt, it proved to be the perfect solution for thousands of small farms, selling more than 322,000 units. *Ralph W. Sanders*

Years of production	1935–1952
Engine	John Deere, 149 cubic inches (2.4 liters), 2 cylinders (until 1938)
	John Deere, 175 cubic inches (2.9 liters), 2 cylinders (1939–1946)
	John Deere, 190 cubic inches (3.1 liters), 2 cylinders (1947–1952)
Fuel	All fuel
Horsepower	16 belt (claimed), 11.8 drawbar (claimed) (1935)
Rated RPM	975 (up to 1938), 1,150 (1939–1946), and 1,250 (1947–1952)
Transmission	4 speeds forward, 1 reverse (1935–1946)
	6 speeds forward, 1 reverse (1947–1952)
Weight	2,760 to 4,000 pounds
Original price	$1,900 (1952)

became the best-selling tractor model produced by Deere during the company's 100-plus years of building tractors.

By the early 1930s, both the A and B were not only available in several specialized versions, but they were the first Deere tractors available with pneumatic rubber tires. While the basic B used a two-wheel tricycle front end, the BN had a single front wheel, the BW had a wide front, the BR had a standard front end, the BO was an orchard model, and the BNH was a high-crop version of the BN. All received larger engines and/or more horsepower in 1940 and 1947, as well as a six-speed transmission instead of the four-speed.

The B was eventually replaced in 1952 by the 50, which used the same 190-cid engine the B used at the end of production.

Allis-Chalmers B

The Allis-Chalmers marketing team that included Harry Merritt, general manager of Allis-Chalmers from 1926 to 1941, had done its homework when it introduced the Model B in 1938. The team figured there were 4 million farmers in the United States who could use farm mechanization, but didn't farm enough acres to justify the purchase of an Allis-Chalmers WC. So, from its inception, the B was designed to be the "successor to the horse." It was designed to use pneumatic tires and replace a team for the low price of just $495.

With about two-thirds the power of the WC, the B was positioned as a one-row, one-plow tractor that was just the right size for any job from cultivating to planting to plowing. The B was even offered with a complete line of mounted attachments that included planters, mowers, and so forth.

While the first 96 B tractors off the line were equipped with 113-cubic-inch Waukesha engines, Allis-Chalmers soon replaced that engine with one of its own design with a 3.25-inch bore and 3.50-inch stroke for 116 cubic inches. In 1943, the stroke was increased to 3.375 inches, increasing the displacement to 125 cubic inches. The rated speed was also increased from 1,400 to 1,500 rpm. As a result, horsepower at the drawbar had climbed to 19.51 when the B was retested on gasoline at the Nebraska Tractor Test Laboratory.

The Allis-Chalmers B is another of those small tractors that found a large market, as evidenced by the sale of more than 127,000 units. One reason is that Harry Merritt, general manager of Allis-Chalmers at the time, figured there were 4 million farmers in the US who could use farm mechanization, but didn't farm enough acres to justify the purchase of an Allis-Chalmers WC. So he insisted that the B was sized to replace a team for the low price of just $495. *Ralph W. Sanders*

One reason the B was so successful was because it was extremely versatile. It not only handled the needs of a small farmer, but also found its way into vegetable operations, nurseries, and a number of non-farm applications. More than 2,800 IB (industrial B model) tractors were built during

The Allis-Chalmers B was just one of several small tractors introduced in the 1930s that were sized and designed for the farmer who was still farming with two or three horses or mules. *Ralph W. Sanders*

Years of production	1938–1957
Engine	Allis-Chalmers, 116.1 cubic inches (1.9 liters), 4 cylinders (until 1943*)
	Allis-Chalmers, 125.2 cubic inches (2.1 liters), 4 cylinders (after 1943)
Fuel	Gasoline
Horsepower	22.25 belt (tested), 19.51 drawbar (tested)
Rated RPM	1,400 (1,500 on later 125-cid engine)
Transmission	3 speeds forward, 1 reverse
Weight	2,620 to 4,193 pounds
Original price	$495 (1938)

* Except for the first 96 B models built, which used a Waukesha 113-cubic-inch (1.9-liter) 4-cylinder engine.

Years of production	1939–1953
Engine	International Harvester, 152.1 cubic inches (2.5 liters), 4 cylinders
Fuel	Gasoline or distillate
Horsepower	26.20 belt (tested), 24.17 drawbar (tested)
Rated RPM	1,650
Transmission	5 speeds forward, 1 reverse
Weight	3,875 pounds
Original price	$750 (1940)

the line's 20-year span. The B line also included specialized units such as the Potato Special and the Asparagus Special, which featured customized axles and clearances.

Starting with serial number 64501, the B could even be equipped with the Allis-Chalmers Snap-Coupler hitch to use with any CA rear-mounted implement. Yet the B could also use a midmount cultivator and midmounted sickle mower that allowed the operator to put the work ahead of him, rather than behind.

By the time the B was discontinued in 1957, the price had gone up to $1,285. Of course, later models included additional features not found on earlier models, such as the hydraulic lift in place of the manual version, and the option of electric lights. With sales totaling more than 127,000 tractors during its 20-year run, the Model B was truly one of the company's most popular tractors—just as management had predicted.

Farmall H

Introduced in 1939 as a replacement for the very successful Farmall F-20, the Farmall H would also become one of the best-selling models in the International Harvester family. Rated as a two-plow tractor, the H featured a roomy platform and a comfortable seat along with individual brake pedals located on the right side of the platform, where they could be locked together.

International Harvester engineers also equipped the H with a new four-cylinder, overhead-valve engine with a 152-cubic-inch displacement. Operating at a faster rated speed of 1,650 rpm, it easily put the H in the 25-horsepower class.

Of course the H also featured the bold new styling developed by famous industrial designer Raymond Loewy. Hence, it featured the bright red paint with silver trim, as well as the contoured sheet-metal hood and new grille that enclosed the radiator.

Despite stiff competition from the John Deere B and other models, like the Case SC and Massey-Harris 101 Junior, the Farmall H was a sales

While one might think the Model M would be the highest-selling model within the Farmall stable of tractors, due to its size and popularity, it is actually the smaller H that holds the record at more than 420,000 units in a number of different versions. *Ralph W. Sanders*

success. In fact, between 1939 and 1954 (the last two model years it was sold as the Super H, which was basically the same tractor with a slightly larger engine and disc-type brakes), the H became the number-two-selling tractor model of all time in North America with 420,011 units built and sold (the last 28,784 being Super H models) in a number of different versions, including an AV cane model.

Case VAC

It's been said that Case marketing wasn't sold on the idea of producing a smaller tractor to compete with popular one-plow models like the John Deere L, Allis-Chalmers B, and Farmall A. They weren't convinced it would

Years of production	1942–1955
Engine	Case, 124 cubic inches (2.0 liters), 4 cylinders
Fuel	Gasoline or distillate
Horsepower	21.33 belt (tested), 19.10 drawbar (tested)
Rated RPM	1,425
Transmission	4 speeds forward, 1 reverse
Weight	2,406 to 4,695 pounds
Original price	$710 (1949 VA with rubber tires)

sell, even though the competitors seemed to be doing quite well. Of course, it was the Case dealers themselves who were seeing the competitors taking sales away from them. Finally, Case management gave in and introduced the V Series with a Continental 124-cubic-inch engine and four-speed transmission.

After two years of production, the V was replaced with the VA Series, which proved to be even more appealing, earning it a place in the Case lineup for several more years. Like its predecessor, it featured a Continental engine, but that was the only major component it shared, even though the appearance was similar. By 1947, Case had developed its own 124-cubic-inch engine, built in the company's Rock Island plant for use in the tractor—replacing even that component.

Had it not been for Case dealers demanding a smaller tractor, the Case VA Series might not have existed. As it turned out, management listened, creating 14 different versions of the tractor in the process, which recorded sales of nearly 150,000 models. *Ralph W. Sanders*

In 1949, the VA models became the first Case tractors to use the new Eagle Hitch and hydraulic lift—the company's answer to the Ford-Ferguson three-point hitch—that allowed snap-on attachment of implements. Like a lot of new row-crop tractors developed at the time, VA Series tractors were available in a number of different versions—14 in this case. The VA was the standard fixed-front-wheel-tread model, while the VAC was the row-crop model with a tricycle front end. The VAC-13 was available for those who wanted an adjustable wide-front model. Other variations included the VAO orchard models, the VAI industrial model, the VAH high-clearance model, and the VAS, which was an offset high-clearance row-crop unit. There was even a VAC-14 that featured a low seat position, which allowed the operator to straddle the transmission, much like the Ford N models.

As it turned out, the VA actually became one of Case's best-selling models, with nearly 150,000 VA Series tractors being sold, including nearly 15,000 industrial models.

Ford 8N

The handshake agreement between Henry Ford and Harry Ferguson had already produced two very successful tractor models—the 9N and 2N—when Henry Ford stepped aside in favor of his grandson, Henry Ford II, in 1945. However, things began to change quickly when the Ford patriarch died in April 1947 at the age of 83. Henry Ford II, and the new team he had hired, immediately saw a problem. Even though Ford was building a product that had become widely popular, the company had no marketing control over the business and was making very little profit off the venture, since the marketing and distribution were handled by Ferguson, per the original agreement.

So, following Henry's death, Ford announced that the handshake agreement was null and void and that the company would stop building tractors for Ferguson and that it would instead be establishing its own marketing and distribution company to handle a new and improved version of the Ford tractor. That, of course, led to a bitter lawsuit that not only cost

Years of production	1947–1952
Engine	Ford, 119 cubic inches (2.0 liters), 4 cylinders
Fuel	Gasoline or distillate
Horsepower	23.56 belt (tested), 16.31 drawbar (tested)
Rated RPM	2,000
Transmission	4 speeds forward, 1 reverse
Weight	2,710 pounds
Original price	$1,404 (1952)

The famous Ford 8N hardly needs an explanation among tractor enthusiasts. As the first tractor built by Ford after Henry Ford's descendants cut ties with Harry Ferguson, the 8N easily outsold the similar-looking rival Ferguson TO-20. In total Ford saw 8N sales top half a million tractors in just five years. *Ralph W. Sanders*

Ford millions of dollars to settle, but millions more in legal fees. The lawsuit also forced Ford to develop a new hydraulic control system and make other changes to avoid using patents that were owned by Ferguson.

Named for the first year of full production, as was the case with the 9N and 2N, the 8N made its debut in late 1947. In an effort to distinguish it from Ferguson's tractors, Ford changed the colors of the 8N from the Ford-Ferguson gray to a red frame and chassis covered by a lighter shade of gray sheet metal. Ford also incorporated a four-speed transmission, for more flexibility, and an improved braking system that positioned both brake pedals on the right side. Ford also raised the steering wheel, installed running boards, and added a position control to the three-point hitch system. A small lever on the right side, under the seat, switched the system between position and draft control. Although the 8N started out with the full Ferguson System that Harry Ferguson had developed, it was eventually modified following the court's decision.

Still, the Ford 8N proved to be a highly successful tractor, outselling by a large margin the similar-looking Ferguson TO-20 that Harry Ferguson released following the split. In the five years that the Ford Ferguson 9N was on the market, the company sold approximately 524,000 units, making it the largest-selling tractor in the United States and Canada during that time period.

Allis-Chalmers WD

When it was introduced in 1948, the Allis-Chalmers WD might have looked similar to the styled WC, but that is where the similarities ended. The late Norm Swinford, who authored *Allis-Chalmers Farm Equipment 1914–1985*, stated in his book, "The new WD was loaded with so many new features and improvements that the sales force had to relearn everything except the name Allis-Chalmers."

The WD not only had approximately 16 percent more power than the WC, but it introduced new features such as two-clutch power control, traction booster, power-shift wheels, and single hitch-point implements. The two-clutch power control was essentially a continuous PTO, which continued to run when the clutch was pushed in. Equally valuable, the single hitch-point implements were unique in that they followed the natural draft lines around curves and over uneven terrain for better depth control. The traction booster, meanwhile, was a system that applied just enough lift on the implement to transfer weight to the tractor to reduce wheel slippage. Key to the special hydraulic system was a new four-piston, high-pressure pump that was driven by four cam lobes on the main driveshaft between the engine and transmission clutches. Although it operated at 3,500 psi, the pump unloaded at the end of the piston stroke to a standby pressure of 1,200 psi for traction boost draft control.

One of the most unique features of the WD tractors was the power-shift rear-wheel adjustment. The patented invention used tractor power to spin the wheels in and out on spiral rails attached to the wheels to change the tread

Years of production	1948–1953
Engine	Allis-Chalmers, 201 cubic inches (3.3 liters), 4 cylinders
Fuel	Distillate or gasoline
Horsepower	34.63 belt (tested), 30.23 drawbar (tested) (gasoline)
Rated RPM	1,400
Transmission	4 speeds forward, 1 reverse
Weight	4,000 pounds
Original price	$1,830 (1948)

The WD was one of the most popular tractors Allis-Chalmers ever built. That was particularly true of tricycle-style tractors, which would accept mounted equipment. *Ralph W. Sanders*

width. A farmer could change the tread width anywhere at any time without the use of jacks, blocks, or any physical effort.

The WD offered more than just innovations. Available in tractor-fuel and gasoline models, it developed up to 34.63 horsepower on the belt and 30.23 horsepower on the drawbar (Nebraska Tractor Test of a gasoline model). Until late 1952, it also had a four-speed transmission, which was upgraded to a constant-mesh transmission in 1952. A year later a battery and distributor replaced the magneto ignition.

Available in dual-front, adjustable-front-axle, single-front-wheel, and cane models, the WD accounted for sales of 146,125 models between 1948 and 1953.

INDEX